v a d e m e c u m

AF896970

Surgical Arithmetic:

Epidemiological, Statistical and Outcome-Based Approach to Surgical Practice

Lawrence Rosenberg, MD, PhD
McGill University
Montreal, Quebec, Canada

Lawrence Joseph, PhD
McGill University
Montreal, Quebec, Canada

Alan Barkun, MD, MSc
McGill University
Montreal, Quebec, Canada

CRC Press
Taylor & Francis Group
Boca Raton London New York

CRC Press is an imprint of the
Taylor & Francis Group, an **informa** business

VADEMECUM
Surgical Arithmetic:
Epidemiological, Statistical and Outcome-Based Approach to Surgical Practice

First published 2000 by Landes Bioscience

Published 2018 by CRC Press
Taylor & Francis Group
6000 Broken Sound Parkway NW, Suite 300
Boca Raton, FL 33487-2742

© 2000 by Taylor & Francis Group, LLC
CRC Press is an imprint of Taylor & Francis Group, an Informa business

No claim to original U.S. Government works

ISBN 13: 978-1-57059-672-8 (pbk)

This book contains information obtained from authentic and highly regarded sources. Reasonable efforts have been made to publish reliable data and information, but the author and publisher cannot assume responsibility for the validity of all materials or the consequences of their use. The authors and publishers have attempted to trace the copyright holders of all material reproduced in this publication and apologize to copyright holders if permission to publish in this form has not been obtained. If any copyright material has not been acknowledged please write and let us know so we may rectify in any future reprint.

Except as permitted under U.S. Copyright Law, no part of this book may be reprinted, reproduced, transmitted, or utilized in any form by any electronic, mechanical, or other means, now known or hereafter invented, including photocopying, microfilming, and recording, or in any information storage or retrieval system, without written permission from the publishers.

For permission to photocopy or use material electronically from this work, please access www. copyright.com (http://www.copyright.com/) or contact the Copyright Clearance Center, Inc. (CCC), 222 Rosewood Drive, Danvers, MA 01923, 978-750-8400. CCC is a not-for-profit organization that provides licenses and registration for a variety of users. For organizations that have been granted a photocopy license by the CCC, a separate system of payment has been arranged.

Trademark Notice: Product or corporate names may be trademarks or registered trademarks, and are used only for identification and explanation without intent to infringe.

Visit the Taylor & Francis Web site at
http://www.taylorandfrancis.com

and the CRC Press Web site at
http://www.crcpress.com

Library of Congress Cataloging-in-Publication Data

Surgical Arithmetic: Epidemiological, Statistical
and Outcome-Based Approach to Surgical Practice / [editors], Lawrence Rosenberg, Lawrence Joseph, Alan Barkun.
p.;cm.--(Vademecum)
Includes bibliographical references and index.
ISBN 1-57059-672-7
1. Surgery--statistical methods. 2. Epidemiology--statistical methods. I. Rosenberg, Lawrence, M.D. II. Joseph, Lawrence, Ph.D. III. Barkun, Alan. IV. Title. V. Series.
[DNLM: 1. Statistics--methods. 2. Surgery. 3. Models, Statistical. WA 950 S961 2000]
RD29.7.S87 2000
617'005'195--dc21 00-057734

While the authors, editors, sponsor and publisher believe that drug selection and dosage and the specifications and usage of equipment and devices, as set forth in this book, are in accord with current recommendations and practice at the time of publication, they make no warranty, expressed or implied, with respect to material described in this book. In view of the ongoing research, equipment development, changes in governmental regulations and the rapid accumulation of information relating to the biomedical sciences, the reader is urged to carefully review and evaluate the information provided herein.

Dedication

To our students and colleagues

Contents

Editors

Lawrence Rosenberg, MD, PhD
Division of Surgical Research
Department of Surgery
McGill University
Montreal, Quebec, Canada
Chapter 9

Lawrence Joseph, PhD
Division of Clinical Epidemiology
Montreal General Hospital
Department of Epidemiology and Biostatistics
McGill University
Montreal, Quebec, Canada
Chapters 2, 3, 7, 8

Alan Barkun, MD, MSc
Division of Gastroenterology
Department of Epidemiology and Biostatistics
McGill University
Montreal, Quebec, Canada
Chapters 1, 3, 8, 12, 13

Contributors

Neena Abraham
Division of Gastroenterology
McGill University Health Centre
Montreal, Quebec, Canada
Chapter 8

Jeffrey Barkun
Divisions of General Surgery, Epidemiology and Biostatistics
McGill University Health Centre
Montreal, Quebec, Canada
Chapter 13

Jean-François Boivin
Centre for Clinical Epidemiology and Community Studies
Sir Mortimer B. Davis Jewish General Hospital
Department of Epidemiology and Biostatistics
McGill University
Montreal, Quebec, Canada
Chapter 1

Ralph Crott
Pharmacoeconomics
Faculty of Pharmacy
University of Montreal
Montreal, Quebec, Canada
Chapter 10

Liane Feldman
Department of General Surgery
Montreal General Hospital
McGill University
Montreal, Quebec, Canada
Chapter 12

Pierre MacNeil
Special Products
Merck Frosst Canada and Co.
Montreal, Quebec, Canada
Chapter 9

Cyr Emile M'Lan
Department of Mathematics and Statistics
McGill University
Montreal, Quebec, Canada
Chapter 6

Robin S. McLeod
Division of General Surgery
Mount Sinai Hospital
Departments of Surgery
and Public Health Sciences
University of Toronto
Toronto, Ontario, Canada
Chapter 4

R. Platt
Montreal Children's Hospital
Research Institute
Department of Epidemiology
and Biostatistics
McGill University
Montreal, Quebec, Canada
Chapter 5

Joseph Romagnuolo
Division of Gastroenterology
McGill University
Montreal, Quebec, Canada
Chapter 3

J. Sampalis
Division of Clinical Epidemiology
Montreal General Hospital
Montreal, Quebec, Canada
Chapter 11

David B. Wolfson
Department of Mathematics
and Statistics
McGill University
Montreal, Quebec, Canada
Chapter 6

Preface

The rapid accumulation of biomedical information and the tremendous pace of technological advancement have combined to make decisions about patient management increasingly difficult for the practicing clinician. The multiplication of competing therapies and the availability of several different technologies for the management of the same condition greatly complicate decision making at both the individual and the policy level. Hence, the need to develop guidelines for clinical practice and to set priorities for healthcare funding is pressing. However, the volume and complexity of medical information thwarts attempts to make firm recommendations about what should and should not be done and what should and should not be paid for.[1]

The overall objective of this book is to meet the growing challenge of making sense of what is known in order to maximize the utility of medical knowledge. Surgeons, in particular, are at the forefront of healthcare professionals that are faced with enormous challenges related to the rapidly changing technology landscape. This book is intended for the practicing surgeon, and hence many of the examples we provide are drawn from the surgical literature. However, it is fair to say that clinicians of every specialty will find it exceedingly useful.

The book is designed to offer practical insights into the essentials of an epidemiological, statistical and outcomes-based approach to surgical practice. Drawing on each of these essential ingredients, surgeons are invited to begin to develop the requisite skills that will allow them to communicate and collaborate effectively with their colleagues in epidemiology and biostatistics in order to develop an optimized approach to patient care—one not only grounded in biomedical science, but in sound evaluative science as well.

We begin with an introduction to epidemiology, the study of disease frequency in human populations. In Chapter 1, *Epidemiologic Methods for Surgery*, Jean-François Boivin and Alan Barkun discuss epidemiologic methods that are useful in addressing a broad range of questions including the determination of the prevalence and incidence of disease, the identification of risk factors, the assessment of prognosis in sick subjects, and the determination of the effectiveness of treatments.

With this foundation, Lawrence Joseph moves on to address basic notions of both descriptive and inferential statistics in Chapter 2, *Introduction to Biostatistics: Describing and Drawing Inferences from Data*. The reader is provided with an explanation of the logic behind common statistical procedures seen in the medical literature, the correct way to interpret the results, and what the drawbacks might be. A special feature is the introduction of Bayesian inference as a strong alternative to standard frequentist statistical methods, both for its ability to incorporate the available prior information into the analysis, and because of its ability to address questions of direct clinical interest.

The interpretation of the literature that is concerned with the evaluation of diagnostic tests and the translation of that information into a form that is easily applicable to a clinical scenario is obviously critical to decision making in clinical practice. In Chapter 3, *Interpretation of Diagnostic Tests*, Joseph Romagnuolo, Alan Barkun and Lawrence Joseph introduce several concepts relating to diagnostic tests and provide examples that demonstrate how these can be used in day-to-day practice.

In Chapter 4, *Primer on Clinical Trials*, Robin McLeod begins to address the question of how one chooses between therapeutic options in her discussion of the randomized clinical trial—often the best trial design for comparing two medical therapies. In this Chapter she discusses the considerations in designing a randomized clinical trial with reference to some of the specia issues in surgical trials and issues related to trial administration.

Univariate statistical techniques, which describe or draw inferences abou the characteristics of a single variable or measurement, are limited since w are often interested in drawing inferences relating to two or more variables or about the effect of one variable on outcome, while adjusting for othe confounding factors. In Chapter 5, *Linear and Logistic Regression Analyses* Robert Platt introduces two primary methods for analysis of multivariat data. These methods allow for the analysis and prediction of continuous o dichotomous outcomes as a function of other variables.

Another broad range of important issues in surgery relates to statistica inference about the probability distribution of the time from some wel defined origin until some well defined event. This is the subject of Chapter 6, *Survival Analysis*, in which David Wolfson and Emil Cyr M'Lan discuss methods that are applicable to many different types of outcomes including "time to remission", "length of postoperative hospital stay", "tim to recurrence of disease", "time free of pain", "time to failure (of a prostheti valve)" and of course "time to death". The Chapter is intended to provide non-technical overview of survival analysis in order to provide a basis fc informed consultation with a biostatistician when the need arises.

Ideally, one should consider all of the available information about an issue before making any clinical decisions. Meta-analysis is helpful in syn thesizing the information about a particular medical issue via statistical analy sis, especially when more than one study has been carried out that relates t the question of interest. In Chapter 7, *A Brief Introduction to Meta-Analysi* Lawrence Joseph provides an overview of the various methods that hav been used in meta-analysis, as well as several of the problems and biases that are difficult to avoid in carrying out a meta-analysis, and how these can be minimized.

Decision analysis attempts to model the possible outcomes from an action. In Chapter 8, *Decision Analysis*, Alan Barkun, Neena Abraham and Lawrence Joseph describe and discuss the basic principles of decision analysis, including decision tree construction, elicitation of probabilities for events in the tree, utility scores for clinical outcomes, cost-effectiveness data and sensitivity analysis to all of these inputs.

In addition to trying to select the best therapy based on effectiveness, in the current era of fiscal restraint and limited resources, it is increasingly important to examine the relationship between effectiveness and cost. In Chapter 9, *Pharmacoeconomics*, Pierre MacNeil and Lawrence Rosenberg provide an introduction to the application of the principles of health economics to the evaluation of therapy. In short, pharmacoeconomics identifies, measures and compares the costs and consequences of use of products and services to the health care system. A structured approach is presented with the aim of providing the clinician a basis for understanding the issues involved.

The discussion of costs, introduced in Chapter 9, is developed in greater detail by Ralph Crott in Chapter 10, *The Costing of Medical Resources*. The purpose of this Chapter is to provide an introduction to how the costs for medical resources are determined and thereby provide the reader with information that can be applied in day-to-day clinical practice. Those dealing with cost-effectiveness analysis, as well as hospital administrators should find this a particularly important Chapter.

Advances in computer technology in recent years have provided the tools by which large databases or registries could be used in clinical research. Disease registries have been implemented, in part, in response to the need for continuous improvement of the health care provided to specific patient populations. Continuous quality improvement requires that data on the treatment and outcome of patients are collected in sufficient detail and continuity in order to allow the evaluation of the impact of interventions. In Chapter 11, *Databases and Registries*, John Sampalis describes the essential requirements and steps in developing a disease registry.

Outcome is defined as a natural result or consequence. Although not stated explicitly, the definition implies that an outcome occurs after some action is taken. In recent years there has been an unprecedented growth in interest in measuring and understanding the outcomes of medical interventions; and surgeons have long been interested in using outcomes to measure quality of care. In Chapter 12, *Do-It-Yourself Outcomes: Using Outcomes in Clinic Practice to Assess and Improve Quality of Care*, Lianne Feldman and Jeff Barku discuss how individual clinicians can design a program in which outcome data are used in everyday clinical practice to improve quality of care. The emphasize that using outcomes information to help answer patients' ques tions about effective treatments is the way of the future—and the futu is now.

Proper use of new and innovative technology requires that as much da as possible is obtained in order to remove uncertainty from the decision making process. In Chapter 13, *Diffusion of Technology*, Alan and Jeffre Barkun bring together many of the issues that were touched upon in prece ing Chapters, including trial design, evaluation of efficacy vs. effectivenes cost, and outcomes. There is little doubt that health care today is drive largely by technology. In this Chapter, the authors examine why new tech nologies need to be evaluated and how new technology is adopted.

In summary, this book introduces a population-based approach to th practice of surgery. It is only through such an orientation that you will b able to make informed decisions about what is best for your patients and f the health care system in which you operate. For many, this will represent new journey of discovery. You will need to learn from the fields of epidem ology and public health, statistics, economics and management. As suc this book represents a complete course on research methods for surgeor Taken as a whole, it offers a comprehensive introduction to all major topi of interest to a clinician who wants to intelligently read the literature or sta his or her own research program.

After the foundations in epidemiology and biostatistics are laid in Cha ters 1 and 2, the remainder of the book can be read in any order. Thus it useful both as a textbook for a comprehensive course on research metho for surgeons and other clinicians, and as a reference on specific topics. N matter what approach you may take, it is our sincere hope that you w come away with practical knowledge about the application of new analytic tools that you can use to assess and improve your own clinical activities a research.

Selected Readings

1. Petitti DB. Introduction. In: Petitti DB, Meta-Analysis, Decision Analysis and C Effectiveness Analysis. New York: Oxford University Press 1994:3-14.

Lawrence Rosenberg, Lawrence Joseph, Alan Bark

Acknowledgments

The authors wish to thank all of the contributors for their outstanding contribution to the completion of this text.

In addition, the authors acknowledge the career support received from the Fords de la Recherchi en Santé chi Québec (FRSQ):

Lawrence Rosenberg is a chercheur-boursier clinicien (senior)

Lawrence Joseph is a chercheur-boursier

Alan Barkun is a chercheur-boursier clinicien

CHAPTER 1

Epidemiologic Methods for Surgery

Jean-François Boivin and Alan Barkun

Epidemiology is the study of disease frequency in human populations. Examples of populations of interest to epidemiologists include the residents of a city, the employees of a hospital, or groups of patients with specific diseases. Epidemiologic methods are useful to address a broad range of questions concerning such populations, including the determination of the prevalence and incidence of disease, the identification of risk factors, the assessment of prognosis in sick subjects, and the determination of the effectiveness of treatments. This Chapter considers applications of epidemiology to surgical practice and research.

Descriptive Epidemiology

Descriptive epidemiology is the study of the distribution of disease frequency in terms of variables of time, place, and characteristics of persons. Table 1.1 gives an example of descriptive statistics concerning the incidence of ectopic pregnancy in the United States.[1] Incidence rates are presented by year, age, and region of the country. An almost fourfold increase in the risk of ectopic pregnancy was seen between 1970 and 1987. The risk increased with age but did not vary much by region of the United States. The authors discussed several hypotheses to explain the observed increase in the rate of diagnosis of ectopic pregnancy over time. One possibility is that frequency of diagnosis of ectopic pregnancy may have increased due to the use of hormone assays for pregnancy confirmation and more sensitive ultrasound imaging. It may also be hypothesized that the increase in risk of ectopic pregnancy was real and not simply due to better diagnostic tests. Various hypotheses about the nature of causal risk factors which may be responsible for this increased frequency may be formulated. The 1970s and 1980s corresponded to a time of changes in contraceptive and sexual practices, which were themselves accompanied by a rise in the incidence of certain gynecologic infections and complications. A specific etiologic hypothesis is that a prior history of salpingitis represents a risk factor for ectopic pregnancy and this question has been addressed by various investigators. Westrom, for example, estimated that ectopic pregnancy occurs 5-10 times more frequently among women with a prior history of salpingitis.[2] Nederlof et al on the other hand, did not find a correlation between ectopic pregnancy rates and salpingitis rates in the United States population, which suggests that other explanations for the epidemic rise in the diagnosis of ectopic pregnancy must be sought.[1] This illustrates how descriptive epidemiologic data may lead to the formulation of specific hypotheses about disease etiology. The converse may also be true: specific causal

Surgical Arithmetic: Epidemiological, Statistical and Outcome-Based Approach to Surgical Practice, by Lawrence Rosenberg, Lawrence Joseph, and Alan Barkun. ©2000 Landes Bioscience.

1

Table 1.1. Rates of ectopic pregnancy by year, age, and region of the United States, 1970-1987[1]

Descriptive variable	Categories	Rate*
Year of diagnosis	1970	4.5
	1975	7.6
	1980	10.5
	1987	16.8
Age (years)	15-24	6.3
	25-34	15.0
	35-44	20.5
Region	Northeast	10.0
	Midwest	10.7
	South	10.8
	West	11.1

*Rate per 1,000 pregnancies (live births, legally induced abortions, and ectopic pregnancies).

factors may be suggested on the basis of other research such as animal work, clinical observations, or other types of epidemiologic studies, and these hypotheses can sometimes be assessed in the light of descriptive information.

Analytic Epidemiology

Most of the research in epidemiology focuses on etiologic or causal hypotheses. A distinction is made between experimental and observational epidemiology. Experimental studies use randomization or similar rigorous treatment allocation procedures to assess the relationship between surgical interventions and outcomes. Marcoux et al, for example, reported a study of laparoscopic surgery in infertile women with minimal or mild endometriosis.[3] The authors determined whether laparoscopic surgery enhanced fecundity in 341 women randomized to either resection or ablation of the visible endometriosis, or to diagnostic laparoscopy only. A 1.9-fold increase in fecundity was observed in the resection-ablation group. The design and analysis of experimental studies represents a specialized topic in epidemiology and this is the subject of Chapter 4 in this book.

Observational epidemiologic studies assess causal relationships in nonexperimental situations where exposure to risk factors happens naturally, outside the context of a planned experimental protocol. Most of clinical practice in medicine is of nonexperimental nature. Several therapeutic interventions cannot be assessed experimentally because of ethical considerations; for example, a blinded randomized trial of laparotomy for a particular gastrointestinal pathology would require sham surgery for the control group, which ethics committees would generally find unacceptable. In other circumstances, treatments should perhaps be assessed experimentally, but resources or interest may be lacking. Similarly, the study of disease etiology and disease prognosis must very frequently rely on observational rather than experimental data.

Cohort Studies

In a cohort study, individuals, all of whom are initially free of the disease under study, are classified according to whether they are exposed or not exposed to the factors of interest. The cohort is then followed for a period of time and the incidence rates in those exposed or not exposed are compared (definition adapted from Kelsey et al.)[4] For instance, Freeman et al reported the results of a cohort study of complications of endoscopic biliary sphincterotomy among 2,347 patients treated at 17 institutions in the United States and Canada.[5] All complications occurred within 30 days of the sphincterotomy and included pancreatitis, hemorrhage, perforation, cholangitis, and cholecystitis. Results for pancreatitis are given in Table 1.2. The term 'exposure variable' or 'risk factor' is used to refer to the factor hypothesized to be associated with an increased or decreased risk of the disease under study. The term 'disease' or 'outcome' may refer to any outcome of interest to the investigator, including handicap, impairment, morbidity, reduced quality of life, and death. In Table 1.2, an increased risk of pancreatitis was observed among patients who, because of difficulties in carrying out the procedure, required a 'precut' sphincterotomy, i.e., an initial cut of the ampulla of Vater in order to facilitate difficult cannulation. The risk of pancreatitis was 0.153 (15.3%) in patients with precut sphincterotomy (17/111) and 0.049 (4.9%), i.e., 110/2,236 in those without precut sphincterotomy. The relative risk, defined as the risk in the exposed divided by the risk in the unexposed, is often calculated to determine the strength of the association between exposure and disease risk. In this example, the relative risk was 0.153/0.049 = 3.1, which means that the risk of pancreatitis was 3.1-fold higher in subjects with precut sphincterotomy than those without. Other measures of association can also be estimated, including the attributable risk or risk difference, i.e., 0.153-0.049 = 0.104. A measure of particular interest is the odds ratio, which is defined as the ratio of the odds of pancreatitis in exposed patients, divided by the odds in unexposed patients. The odds of disease in the exposed are estimated as the number of subjects with pancreatitis divided by the number without, i.e., 17/94. Similarly, the odds in the unexposed are 110/2,126, and the odds ratio is then (17/94)/(110/2,126) = (17x2,126)/(94x110) = 3.5. It can be seen that the estimate of the odds ratio, i.e., 3.5, resembles the estimate of the relative risk obtained earlier, i.e., 3.1. The odds ratio represents a good approximation of the relative risk when the risk of the disease under study, i.e., pancreatitis in the present example, is low, say less than 10%. We have seen that this risk was actually 15.3% in the exposed subjects, which explains why in this example the odds ratio and the relative risk are somewhat different, albeit not markedly. This approximate relationship between the odds ratio and the relative risk is useful in certain epidemiologic contexts, for example in case-control studies of short-term effects, which we will now discuss.

Table 1.2. A cohort study of pancreatitis in patients with sphincterotomy[5]

	Precut sphincterotomy (exposure = yes)	No precut sphincterotomy (exposure = no)	Total
Pancreatitis (Outcome = Yes)	17 (15.3%)	110 (4.9%)	127
No pancreatitis (Outcome = No)	94 (84.7%)	2,126 (95.1%)	2,220
Total	111 (100%)	2,236 (100%)	2,347

Relative risk: (17/111) + (110/2,236) = 3.1

Odds ratio: (17/94) + (110/2,126) = 3.5

Case-Control Studies

The sphincterotomy study discussed above was called a cohort study because a group of subjects exposed to precut sphincterotomy was compared to a group of unexposed subjects; these exposed and unexposed subjects formed a cohort in which risks of disease were assessed and compared. An alternative strategy to determine complications of sphincterotomy could have been to find cases of pancreatitis and a certain number of controls without pancreatitis. The investigators could have for example decided to include all possible cases of pancreatitis available in the collaborating institutions, and an equal number of controls. Expected results from this study design are presented in Table 1.3. Because the decision about the number of cases and controls to be included in the study is made by the investigators, the calculation of risks in the exposed and the unexposed subjects becomes inappropriate. For example, the observed risk of pancreatitis in the exposed subjects in Table 1.2 was 15.3% (17/111); if we use the Table 1.3 data, we obtain the incorrect value of 77.3% (17/22 = WRONG!). More complex case-control designs may allow the estimation of risks,[6] but this topic is outside the scope of this book. Since risks cannot generally be estimated validly in case-control studies, relative risks and attributable risks are also unavailable. It turns out, however, that the odds ratio can be estimated validly. In Table 1.3, for example, the odds ratio is (17x122)/(110x5)=3.7, identical (except for rounding) to the value estimated in the cohort data presented in Table 1.2. Since the odds ratio calculated in Table 1.2 represented an approximation to the relative risk, and since this odds ratio is also available from case-control studies, case-control studies can provide an approximation to the relative risk, even if the risks themselves cannot be calculated. A more formal demonstration of these relationships can be found in various specialized textbooks.[7]

Table 1.4 presents an example of a case-control study of HIV seroconversion in health care workers after possible percutaneous exposure to the virus.[8] Cases and controls were all health care workers who had an occupational percutaneous exposure to HIV-infected blood. The investigators assessed factors which influenced the risk of HIV infection. One of the exposure variables was the severity of the injury and Table 1.4 shows results concerning exposure to deep punctures or wounds. As

Table 1.3. A case-control study of pancreatitis in patients with sphincterotomy (fictitious example)

	Precut sphincterotomy (exposure = yes)	No precut sphincterotomy (exposure = no)	Total
Pancreatitis (Outcome = Yes)	17 (13.4%)	110 (86.6%)	127(100%)
No pancreatitis (Outcome = No)	5 (3.9%)	122 (96.1%)	127 (100%)
Total	22	232	254

Odds ratio: (17 x 122) + (110 x 5) = 3.8

Table 1.4. A case-control study of HIV seroconversion in health care workers after percutaneous exposure[8]

	Deep injury (exposure = yes)	No deep injury (exposure = no)	Total
HIV seroconversion (Outcome = Yes)	17 (51.5%)	16 (48.5%)	33 (100%)
No HIV seroconversion (Outcome = No)	46 (6.8%)	629 (93.2%)	675 (100%)

Odds ratio: (17 x 629) + (16 x 46) = 14.5

explained above, it is not possible to estimate the risks of seroconversion in workers exposed to deep injury and those not exposed, but an approximate estimate of the relative risk can be obtained through the odds ratio, as (17 x 629)/(16 x 46) = 14.5. This means that the risk of seroconversion was 14.5 higher in workers who sustained a deep injury at the time of percutaneous exposure than among workers who sustained a more superficial injury.

Person-Time

The examples discussed above did not involve long intervals of time between exposure to the risk factor and occurrence of the disease of interest. In the study of pancreatitis after sphincterotomy, all cases of pancreatitis were diagnosed within 30 days of the surgical intervention. In the HIV seroconversion study, we can also presume that the relevant time window was short. In other circumstances, however, the analysis of the scientific question of interest requires a long time interval between exposure and outcome. Table 1.5 gives an example. Shibata et al assessed the relationship between the DCC (deleted in colorectal cancer) protein and prognosis of colorectal cancer.[9] The investigators determined the DCC status of 132 patients treated for colorectal cancer between 1965 and 1990 and they followed these patients

1

Table 1.5. A cohort study of prognosis in patients with colorectal cancer[9]

	DCC status negative (exposure = yes)	DCC status positive (exposure = no)	Total
Death	44	24	68
Person-months at risk	66 patients x 85.1 mos of average follow-up = 5,167 person-months	66 patients x 95.7 mos of average follow-up = 6,316 person-months	11,483 person-months

Mortality rate ratio: (44 deaths/5,167 person-months)/(24 deaths/6,316 person/months) = 2.2

until 1996 to determine their survival status. In this type of research, long-term results are of interest, and the patients were followed for an average of 92.1 months. The duration of follow-up differed markedly between subjects, depending on their date of initial treatment for colorectal cancer and their date of death, and a few subjects were followed for more than 20 years after initial treatment. In addition, the average follow-up time was shorter in the DCC negative group, in which mortality was higher. Because of all of these temporal features which differed from patient to patient, the study design and analysis must take time into account. Table 1.5 presents results of the colorectal study in terms of person-time at risk. Each study subject contributed to the denominator of the incidence rates the amount of time between his/her initial treatment for colorectal cancer and the end of his/her follow-up, which may be his/her date of death or simply the date of the end of the study in 1996; subjects lost to follow-up also contributed the time during which their status was known to the investigators. The mortality rates were 44 deaths/5,167 person-months = 852 deaths per 100,000 person-months in the DCC negative group, and 24 deaths/6,316 person-months = 380 deaths per 100,000 person-months among the positive patients. The mortality rate ratio was therefore (852/100,000)/(380/100,000) = 2.2, and the mortality rate difference was (852/100,000)-(380/100,000) = 472/100,000.

Case-control studies may also be used to investigate associations requiring relatively long follow-up intervals for the assessment of risk. Table 1.6 presents results of a case-control study of the association between appetite-suppressant drugs and primary pulmonary hypertension.[10] Ninety-five patients with primary pulmonary hypertension and 335 controls without this disease were identified in four European countries. Primary pulmonary hypertension may occur a long time after exposure, and the concept of person-time at risk is therefore applicable to the study of this scientific question. The use of appetite suppressants was determined for all subjects included in the study. Table 1.6 shows that 31.6% of the cases (30/95) and 7.3% of the controls (26/355), respectively, used these drugs. Because of the case-control design, incidence rates of primary pulmonary hypertension in the exposed and in the unexposed subjects cannot be calculated (see the earlier discussion about the unavailability of risk estimates risks in case-control studies). It can be shown, however, that if controls are selected appropriately, the odds ratio from such case-controls studies represents an exact estimate of the incidence rate ratio which would have been obtained, had a cohort study been conducted rather than a case-control study.[11] In the example in Table 1.6, the incidence rate ratio is therefore 5.8 (=(30x329)/(65x26)). This means that the incidence rate of primary pulmonary hypertension is 5.8 times larger in subjects exposed to appetite suppressants than in unexposed subjects. A demonstration of the link between the odds ratio and the incidence rate ratio in case-control studies is beyond the scope of this Chapter and the reader is referred to appropriate references.[7]

An example of a case-control study of risk factors for a disease of surgical interest is the report published by Neugut et al[12] on the association between cholecystectomy and the occurrence of colon cancer.

Interpretation of Results

In several of the examples discussed until now, causality questions were raised. Investigators were interested in whether a given exposure, precut sphincterotomy, deep injury with HIV-contaminated material, DCC protein negativity, or appetite suppressant drug use, caused morbidity or mortality. The data presented in Tables 1.2, 1.4, 1.5, and 1.6 show simple univariate relationships between exposure and outcome, and all associations were positive, indicating an increased frequency of infection, disease, or death after exposure. The presence of a positive association suggests that a causal link may be present, but several additional considerations must be taken into account before forming a judgment about causality. These include temporal sequence, precision, and bias.

Temporal Sequence

A foremost issue in epidemiologic studies is the question of the temporal relationship between exposure and outcome. Did the exposure under study precede the occurrence of disease or did it possibly occur after disease had developed? It is conceptually obvious that to be causal, exposure must precede disease. In practice, however, the actual temporal sequence of events is not always easy to disentangle. Table 1.7 shows results of a study of the presence of *Chlamydia* species in coronary artery specimens of 90 symptomatic patients undergoing coronary atherectomy and of 24 control patients without atherosclerosis.[13] The coronary specimens of these subjects were tested for the presence of *Chlamydia* species using direct immunofluorescence. The odds ratio was extremely large ((71x23)/(19x1) = 85.9), indicating a strong positive association between the presence of *Chlamydia* and symptomatic atherosclerotic cardiovascular disease. The authors concluded that *Chlamydia* infection may have a causal role in the development of coronary atherosclerosis. In their discussion, however, the authors indicated that their study design could not distinguish between the situation where *Chlamydia* infection led to atherosclerosis, and the situation where it was the atherosclerotic plaques which represented more fertile grounds for the *Chlamydia* to be deposited and grow. If this were the case, the presence of the pathogens would be a result rather than a cause of the disease. A clarification of the actual temporal sequence would require a different study design in which the presence of infection would be determined before the incidence of disease. In some epidemiologic studies, the temporal sequence of events is clear. In the case-control study of HIV infection in health care workers discussed earlier, for

Table 1.6. A case-control study of appetite suppressant drugs and primary pulmonary hypertension[10]

	Use of appetite suppressants (exposure = yes)	**No use of appetite suppressants (exposure = no)**	**Total**
Primary pulmonary hypertension (Outcome = Yes)	30 (31.6%)	65 (68.4%)	95 (100%)
No primary pulmonary hypertension (Outcome = No)	26 (7.3%)	329 (92.7%)	355 (100%)

Odds ratio: (30 x 329) + (65 x 26) = 5.8

Table 1.7. A case-control study of* Chlamydia *infection and coronary atherosclerosis[13]

	Chlamydia present (exposure = yes)	**Chlamydia absent (exposure = no)**	**Total**
Atherosclerotic coronary disease (Outcome = Yes)	71 (78.9%)	19 (21.1%)	90 (100%)
No atherosclerotic coronary disease (Outcome = No)	1 (4.2%)	23 (95.8%)	24 (100%)

Odds ratio: (71 x 23) + (19 x 1) = 85.9

example, the authors only included subjects who were documented as being seronegative at the time of traumatic exposure to HIV-contaminated material.[8] Similarly, in the cohort study of DCC protein status, the outcome was death, an event which was clearly separated in time from the time of assessment of DCC protein status.[9] In other studies, the situation may be somewhat ambiguous or in the extreme, such as in the case of the *Chlamydia* infection study, totally unclear. An adequate assessment of an epidemiologic study must always therefore include as an essential element a consideration of whether the temporal sequence was adequately characterized.

Precision

Subjects included in epidemiologic studies can generally be seen as a sample from a much larger, possibly infinite, population. This population may be conceptual (for example, all future cases of HIV infection) or real (for example, all actual cases on a given territory). For example, in the case-control study of primary pulmonary hypertension presented earlier, 95 cases of the disease were included while thousands of such cases are actually known.[10] The estimate of the relative risk can therefore be considered as somewhat imprecise, being based on observations from a limited subset of the entire population of such cases. The assessment of precision of the estimates derived from epidemiologic studies relies on statistical methodology. Statistical issues in epidemiologic research are covered in Chapter 2 of this book.

1

Bias

The validity of results from an epidemiologic study may be affected for several reasons. Study subjects may not be comparable, with for example older patients among the exposed members of a study than among the unexposed. The method of selection of the study subjects may be biased in such a way as to lead for example to an exaggerated level of exposure in cases, and therefore an inflated odds ratio. Various errors may occur in the collection and recording of the data, leading to erroneous estimates of the measures of association. A standard approach in classifying these various types of biases is to regroup them under the categories of confounding, selection bias, and information bias.

Confounding

Confounding arises when exposed and unexposed subjects are not comparable for a variable which represents a risk factor for the outcome of interest. Tables 1.8 and 1.9 give an example. In Table 1.8, results of a cohort study of the association between radiotherapy to the chest for Hodgkin's disease and subsequent mortality from coronary artery disease are presented.[14] The risk of coronary disease death was 2.0% in the exposed (70/3530), and 4.0% in the unexposed (54/1360). The relative risk was therefore 0.5 (= 2%/4%), indicating a reduced coronary disease mortality after radiation to the chest. This result was surprising, as the investigators' hypothesis, arising from previous research in other types of patients and also from animal models, was that exposure to radiation would increase, and not decrease, coronary disease mortality. Further analysis of these data revealed, however, that these results were biased because of the confounding effect of age. Table 1.9 presents results of the same study, stratified into three age groups. The relative risk of coronary disease death was estimated separately for each of these three age groups, and the direction of the association was now positive, with coronary disease mortality being larger in the exposed subjects than in the unexposed subjects in each of the three groups. The difference between the relative risk estimated from the data presented in Table 1.8 and those presented in Table 1.9 is due to a confounding bias arising from the unequal distribution of the exposed and unexposed subjects with respect to age. We can see in Table 1.9 that 76% (2,680/3,530) of the exposed subjects were 0-39 years old while only 35% (470/1,360) of the unexposed subjects were in the same age category. The exposed population was therefore much younger than the unexposed, and since coronary disease mortality is known to be lower in younger subjects, the comparison of exposed and unexposed subjects was biased by age. When results are considered within age strata, this bias is corrected, and this is why the relative risks in Table 1.9 show an increased risk rather than a decreased risk as suggested by the biased comparison based on Table 1.8. The effect of confounding variables may be

1

Table 1.8. A cohort study of coronary artery disease mortality in patients treated with radiation for Hodgkin's disease (adapted from Boivin et al)*[14]

	Radiotherapy to the chest (exposure = yes)	**No radiotherapy to the chest (exposure = no)**
Coronary disease death (Outcome = Yes)	70 (2%)	54 (4%)
No coronary disease death (Outcome = No)	3,460 (98%)	1,306 (96%)
Total	3,530 (100%)	1,360 (100%)

*The study conducted by Boivin et al was a case-cohort study, an epidemiologic study design not presented in this book. To facilitate the presentation of this example, the reported data were modified to correspond to a cohort design.
Crude relative risk: (70/3,530) + (54/1,360) = 0.5

Table 1.9. A cohort study of coronary artery disease mortality in patients treated with radiation for Hodgkin's disease: results stratified by age (adapted from Boivin et al[14])

Coronary disease death	**Radiotherapy to the chest (exposure = yes)**	**No radiotherapy to the chest (exposure = no)**	**Relative risk**
AGE = 0-39 yr			
Yes	21	2	
No	2,659	468	1.8
Total	2,680	470	
Age = 40-59 yr			
Yes	26	13	
No	704	437	1.2
Total	730	450	
Age = 60+ yr			
Yes	23	39	
No	97	401	2.2
Total	120	440	
Grand Total, All Ages	**3,530**	**1,360**	

controlled using various strategies such as matching in the selection of study subjects, or multivariate techniques in the analysis of the data. Multivariate analysis is discussed in Chapter 5 of this book.

Selection Bias

Results of epidemiologic studies may also be biased because of inappropriate selection of study subjects. One type of selection bias is called referral bias. Referral bias may arise when procedures used to identify disease status vary with exposure. We give here an example in the context of studies of the relationship between oral contraceptives and thrombophlebitis. The medical literature suggests that oral contraceptives can cause thrombophlebitis. Therefore, if a woman consulting her family physician in a community clinic has signs or symptoms of thrombophlebitis, the physician should inquire about oral contraceptive use. If the woman does use oral contraceptives, then the physician will be more readily convinced that the correct diagnosis is thrombophlebitis and he/she will then refer the woman to the hospital for a thorough diagnostic assessment of thrombophlebitis. If, however, the woman does not use oral contraceptives, the physician may think that the diagnosis of thrombophlebitis is less likely and may decide to simply send the woman home, perhaps with the advice of using analgesics. In some of these women, the diagnosis of thrombophlebitis may have been missed and the condition may improve spontaneously, without further treatment. Now, if we conduct a study of hospitalized patients with thrombophlebitis, we may obtain a larger estimate of effect of oral contraceptive use and risk thrombophlebitis than the true value. This is because there was more of a tendency to refer women with thrombophlebitis who use oral contraceptives to the hospital than similar women who do not use contraceptives.

Several other types of selection bias exist, affecting both cohort and case-control studies. The reader is referred to specialized textbooks for a more systematic review of this question.[4,6,7]

Observation Bias

Observation bias refers to errors in the classification of study subjects with respect to study variables. A certain number of exposed subjects in a cohort or a case-control study may for example be incorrectly classified as unexposed, or vice-versa; similarly, cases of disease may be incorrectly classified as noncases, and noncases as cases. Such errors may affect the estimates of the measures of association such as the relative risk or the odds ratio, and the impact will vary depending on the pattern of these errors. When the errors are nonsystematic, i.e., when they affect exposed and nonexposed equally, or cases and noncases equally, it has been shown that the impact of misclassification will generally be in the direction of attenuating the strength of association, or, in other words, of producing measures of association which are smaller than the true values. When the errors are systematic, the impact may be in any direction, depending on the distribution of these errors. An example of a systematic error would be to tend to classify people who complain of depressive symptoms as clinically depressed when they have a family history of depression, and to do this more often than when they have no such history. In such a situation, the association between family history could be falsely exaggerated. Another example is the study presented in Table 1.6, in which Abenhaim et al assessed the use of appetite

1

suppressants in cases of primary pulmonary hypertension and in noncases.[10] Each subject underwent a thorough, face-to-face interview about exposure to drugs. Patients with primary pulmonary hypertension might have been more likely to remember using anorexic agents than controls. If this were true, controls would sometimes be classified as unexposed while they were actually exposed, while cases would tend to provide correct exposure information. The impact would be to inflate inappropriately the odds ratio. Abenhaim et al verified with drug sales figures the likelihood of such errors, and recall bias seemed improbable.[10] Rothman and Greenland[7] provide a systematic review of misclassification.

Conclusion

Several considerations come into play when selecting a study design for an epidemiologic study, and no general rule is applicable to all situations. Some general principles may, however, be formulated. When the exposure under study is rare, a cohort design may be preferable. For example, very few people in the general population of a country or of a city have had an endoscopic biliary sphincterotomy, with or without a precut technique. In planning such a study, investigators will want to be certain that they will recruit a sufficient number of exposed subjects. The cohort design selected by Freeman et al, in which exposed and nonexposed subjects were recruited from 17 health care institutions, allowed them to identify a sufficient number of subjects with the treatments of interest.[5] An additional advantage of the cohort design selected by Freeman et al was that they could also investigate the risk of several other outcomes such as hemorrhage, cholangitis, etc. On the other hand, case-control designs are useful when the disease under study is rare. In the study of HIV seroconversion in health care workers, for example, both the exposure and the disease were rare in the general population, and the case-control design made it possible to identify a reasonable number of cases without having to recruit a huge cohort of health care workers.[8] The case-control design is also sometimes selected when several risk factors are investigated. In the HIV seroconversion study, for example, the type of device causing the injury, the use of gloves, the use of zidovudine after the injury, etc., were also investigated. Hybrid study designs, such as nested case-control studies and case-cohort studies, which combine features of both cohort and case-control studies, have also been developed. These more advanced designs are described elsewhere.[4,6,7]

Selected Readings

1. Nederlof KP, Lawson HW, Saftlas AF et al. Ectopic pregnancy surveillance, United States, 1970-1987. Morbidity and Mortality Weekly Report 1990; 39 (No SS-4):9-17.
2. Weström L. Influence of sexually transmitted diseases on sterility and ectopic pregnancy. Acta Europaea Fertilitatis 1985; 16:21-24.
3. Marcoux S, Maheux R, Bérubé S, and the Canadian Collaborative Group on Endometriosis. Laparoscopic surgery in infertile women with minimal or mild endometriosis. N Eng J Med 1997; 337:217-222.
4. Kelsey JL, Whittemore AS, Evans AS et al. Methods in observational epidemiology. Second edition. New York: Oxford University Press. 1996.
5. Freeman ML, Nelson DB, Sherman S et al. Complications of endoscopic biliary sphincterotomy. N Eng J Med 1996; 335:909-918.

6. MacMahon B, Trichopoulos D. Epidemiology. Principles and methods. Second edition. Boston: Little, Brown and Company. 1996.
7. Rothman KJ, Greenland S. Modern epidemiology. Second edition. Philadelphia: Lippincott-Raven Publishers. 1998.
8. Cardo DM, Culver DH, Ciesielski CA et al. A case-control study of HIV seroconversion in health care workers after percutaneous exposure. New England Journal of Medicine 1997; 337:1485-1490.
9. Shibata D, Reale MA, Lavin P et al. The DCC protein and prognosis in colorectal cancer. N Eng J Med 1996; 335:1727-1732.
10. Abenhaim L, Moride Y, Brenot F et al. Appetite-suppressant drugs and the risk of primary pulmonary hypertension. N Eng J Med 1996; 335:609-616.
11. Miettinen OS. Estimability and estimation in case-referent studies. Amer J Epidemiol 1976; 103:226-235.
12. Neugut AI, Murray TI, Garbowski GC et al. Cholecystectomy as a risk factor for colorectal adenomatous polyps and carcinoma. Cancer 1991; 68:1644-1647.
13. Muhlestein JB, Hammond E, Carlquist JF et al. Increased incidence of *Chlamydia* species within coronary arteries of patients with symptomatic atherosclerosis versus other forms of cardiovascular disease. J Amer Coll Cardiol 1996; 27:1555-1561.
14. Boivin JF, Hutchison GB, Lubin JH et al. Coronary artery disease mortality in patients treated for Hodgkin's disease. Cancer 1992; 69:1241-1247.

CHAPTER 2

Introduction to Biostatistics: Describing and Drawing Inferences from Data

Lawrence Joseph

I. Introduction

Consider the following statements, which were included as part of an abstract to an article reporting the results from a randomized trial comparing stent placements to balloon angiography in obstructed coronary bypass grafts:

> "As compared with the patients assigned to angioplasty, those assigned to stenting had a higher rate of procedural efficacy...(92% vs. 69%, $p < 0.001$), but they had more frequent hemorrhagic complications (17% vs. 5%, $p < 0.01$).... The outcome in terms of freedom from death, myocardial infarction, repeated bypass surgery, or revascularization of the target lesion was significantly better in the stent group (73% vs. 58%, $p = 0.03$)"[1]

Proper interpretation of the above results, and of similar reports from much of the modern clinical literature, depend in large part on the understanding of statistical terms. In this case, terms such as "significant" and "*p*-values" were given, and in other reports one may see terms like "confidence intervals", "*t* tests", "Chi-squared tests", "power", "type I and type II errors", and so on. Clearly, surgeons and other clinicians who wish to keep pace with new techniques and technologies must at least have a basic understanding of statistical language. This is true not only if they desire to plan and carry out their own research, but also if they simply want to read the medical literature with a keen critical eye, or if they want to make informed decisions about which new treatments they may wish to apply to their own patients, and under which circumstances.

This Chapter will introduce the basic notions of both descriptive and inferential statistics. A distinguishing feature of our approach will be to explain in some detail the inferential ideas behind the most commonly used statistical tests and other techniques. Therefore, rather than simply providing a catalogue of which formulae to use in which situation, we also provide the logic behind each technique. In this way, informed choices and decisions can be made, based on a deeper understanding of exactly what information each type of statistical inference provides.

Section 2 will cover simple graphical and numerical techniques for describing data. Section 3 presents a brief introduction to rules of probability which underlie all inferential techniques. Basic ideas of statistical inference are introduced in

Surgical Arithmetic: Epidemiological, Statistical and Outcome-Based Approach to Surgical Practice, by Lawrence Rosenberg, Lawrence Joseph, and Alan Barkun. ©2000 Landes Bioscience.

Section 4, and applied in Sections 5 and 6 to problems involving proportions and means, respectively. Section 7 briefly discusses nonparametric techniques, and introduces the correlation between two variables. In these sections, we will learn exactly what is meant by ubiquitous statistical statements such as "$p < 0.05$" (which may not mean what many medical journal readers think it means!), and examine confidence intervals as an attractive alternative to p-values. The problem of choosing an appropriate sample size for a given experiment is discussed in Section 8. Increasingly important Bayesian alternatives to "classical" or "standard" statistical techniques are presented in Section 9, and we conclude with a summary and brief reference list for further reading.

2

2. Descriptive Statistics

The first step in most statistical analyses is to examine the data at hand using simple graphical and numerical summaries. These descriptions will often be of interest in themselves, and are also helpful for selecting the most appropriate inferential techniques to be used later. Another important use of descriptive statistics is to find unusual or even impossible values (for example, values that are outside of the feasible range) in the data set, that may either arise from errors in data entry or from subjects whose profiles are very different from those of others (outliers).

Generally speaking, graphical summaries are useful for presenting overall features of a data set, while numerical summaries provide more exact descriptions of specific features. Below we discuss the most common descriptive statistics that are found in the medical literature. We will illustrate the techniques using data that came from a randomized controlled trial of laparoscopic versus open surgery for hernia repair. For ease of illustration, we will focus on a subset of the data from 25 subjects, including information about the surgical group the patient was randomized to, the patients' gender, number of years smoking, and a major outcome of the study, days to convalescence. The subset of data from this trial are given in Table 2.1. See Chapter 5 for the complete data set and more on its analysis.

From Table 2.1, we first notice that there are different types of data variables. The type of data in the column headed "Group" is called discrete data, since the possible values for the data come in discrete steps. In fact, in this case there are only two possible values, a 0 denoting conventional surgery, and a 1 denoting laparoscopic surgery. When discrete variables can have only two values, they are often referred to as dichotomous data. Sex, for example, is also a dichotomous variable. Days of convalescence, on the other hand, can take on all possible values from 0 and higher. These types of variables are usually referred to as continuous variables, since they may take on all possible values within their range, at least in theory. In practice, the way variables are measured or recorded mean that almost all continuous variables are recorded as discrete. For example, while in theory convalescence time can take on any value greater than 0 and we age continuously, we often record convalescence time in full days and age as an integer number of years. Nevertheless, it is convenient to consider continuous variables as representing an underlying continuum whenever the number of possible values that are recorded is large.

2

Table 2.1. Data from 25 subjects participating in a randomized clinical trial of conventional versus laparoscopic surgery for hernia repair.

Patient #	Group	Sex	Smoke Years	Days of Convalescence
1	0	1	0	21
2	1	1	0	4
3	1	1	0	3
4	0	1	0	12
5	1	1	12	4
6	1	1	0	5
7	0	1	9	11
8	1	1	0	20
9	0	0	0	28
10	0	1	0	3
11	1	1	0	22
12	1	1	0	5
13	0	1	0	NA
14	0	1	0	10
15	1	1	0	12
16	1	1	30	15
17	0	1	40	9
18	1	1	0	5
19	1	1	0	1
20	0	1	30	5
21	1	1	10	14
22	0	1	0	7
23	0	1	0	10
24	1	1	50	13
25	0	1	40	6

Group 0 received conventional surgery, while Group 1 received laparoscopic surgery. Females are indicated by a 0, while males are denoted by a 1. Smoke years is defined as the number of years smoking for current smokers, and 0 otherwise. NA denotes "not available".

2.1. Graphical Summaries

Figure 2.1 provides a histogram of the number of days to convalescence for this group of 25 patients. Histograms are constructed by breaking up the range of the variable of interest into disjoint intervals of (usually) equal width, and displaying the count or proportion of observations falling into each interval. From this histogram, we can observe that most subjects had less than 15 days of convalescence, although there was, for example, one patient with 25-30 days, and another two subjects with 20-25 days to convalescence. This histogram presents nonsymmetrical or "skewed" data.

While histograms are very common and are useful for displaying the overall shape and range of a variable in a data set, boxplots are often more convenient for comparing two groups. Figure 2.2 presents two boxplots of the convalescence days by surgery groups. The upper and lower limits in the boxplot indicate the maximum and minimum values in the data set, while the median line is in the middle. The

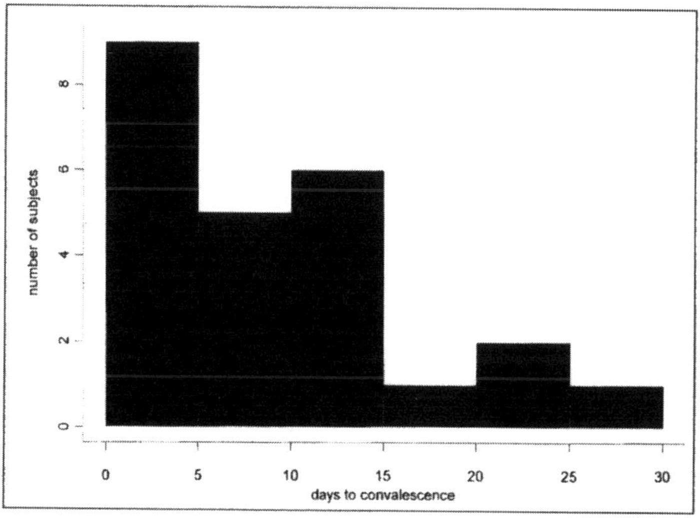

Fig. 2.1. Histogram of the numbers of days of convalescence, using the data given in the last column of Table 1.1.

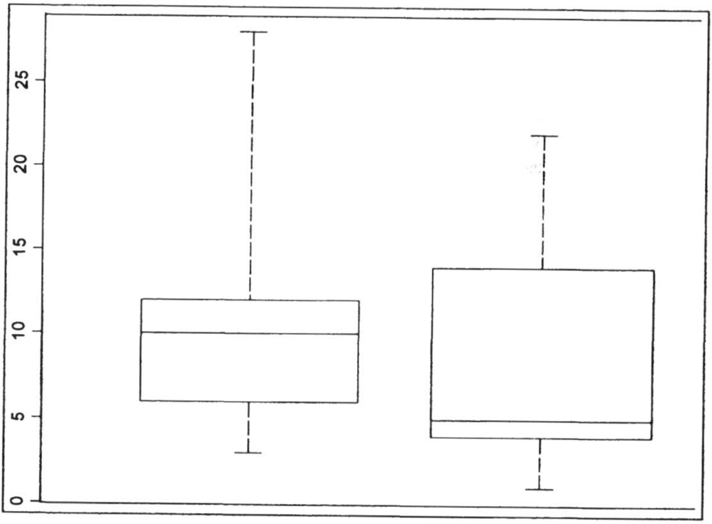

Fig. 2.2. Boxplots of the numbers of days of convalescence in the conventional (on the left) and laparascopic groups.

median is the value such that exactly half of the data values lie above and half lie below the line. From Figure 2.2, we see that the median value in the conventional surgery group (Group 0) is approximately 10, while the median number of days to convalescence is only 5 days in those who received laparoscopic surgery (Group 1). The upper and lower limits of the boxes indicate the upper and lower quartiles of the samples, so that, for example, about 25% of subjects in Group 0 had values of 12 or more days, while 25% had values of 6 or below. The interval enclosed in the box, here 6-12 days for Group 0 and 4-13 days for Group 1, is often called the interquartile range of a sample.

There are many other types of graphics that are used, including stem-and-leaf plots and scatter plots. A stem and leaf plot resembles a histogram turned on its side, and is especially useful for small data sets where it may be of interest to see individual data values. In a stem and leaf plot, the first column lists the initial digits of each number, while the final digit is used to form the histogram. For example, from the stem-and-leaf plot in Figure 2.3, we can see that the lowest value is 1 (a zero on the left and a 1 on the right of the colons), the next two lowest values are both equal to 3, and so on. The maximum value is seen to be 28, while from the histogram in Figure 2.1 we knew only that it was between 25 and 30.

While the graphics so far have looked only at one variable at a time, a scatter plot is useful for examining the relationship between two variables. Figure 2.4 presents a scatter plot of the relationship between the number of years smoking and the number of days to convalescence. There is no obvious relationship between the two variables in this figure, since the values of one do not tend to increase or decrease as the other variable moves through its range. More formal statistical procedures for assessing the effects of one variable on another will be found in Chapter 5, although we will briefly look at the correlation between two variables in Section 7 of this Chapter.

2.2. Numerical Summaries

Numerical summaries can complement the information found in graphs such as those presented above. The most common numerical summaries are defined below:

Mean

The mean of a set of numbers is the usual numerical average, found by summing all the numbers and dividing by the sample size. For example, the mean number of days to convalescence for the 25 subjects in Table 2.1 is

$$(21 + 4 + 3 + \cdots + 6)/24 = 10.21 \text{ days.}$$

Note that in order to calculate this average, we had to delete patient #13, whose value for days to convalescence was missing, so that the divisor was 24 rather than 25.

Median

As mentioned above, the median is the number such that half of the values in the data set are equal to or above the median, and half are equal to or below the median. For the convalescence data in Table 2.1, the median number could be chosen to be any number between 9 and 10, although the midpoint of 9.5 is usually chosen. Note that there are 12 values above 9.5 and 12 values below 9.5 in this data set.

```
0 : 13344
0 : 5555679
1 : 0012234
1 : 5
2 : 012
2 : 8
```

Fig. 2.3. A stem-and-leaf plot for days of convalescence, using the data from the last column of Table 2.1.)

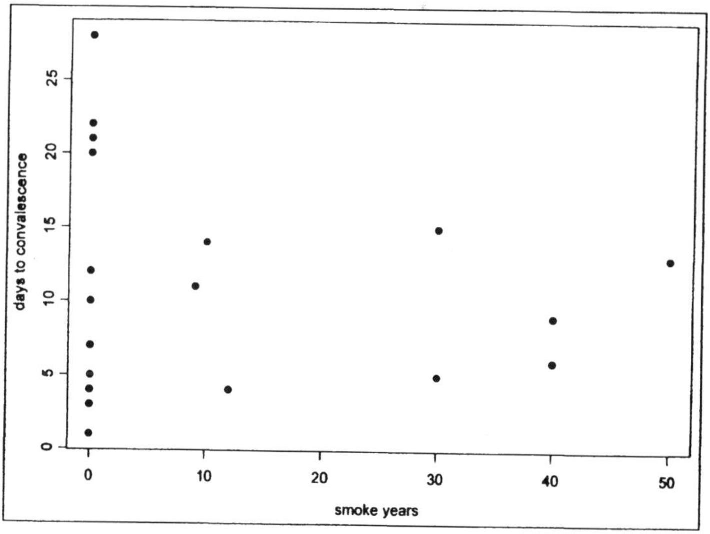

Fig. 2.4. Scatter plot of smoke years versus days of convalescence.

While both the median and the mean are measures of central tendency, the median tends to be more representative of central tendencies in skewed data sets compared to the mean. For example, suppose that instead of 28 being the largest time to convalescence in the data in Table 2.1, it was 280. The mean value would be largely affected by this, more than doubling its' value from 10.21 to 20.71, while the median remains unchanged at 9.5.

Variance and Standard Deviation

The variance is the average squared distance of each data point to the mean of the sample. It is calculated by first finding the mean, next summing the square of each data point minus that mean, and finally dividing by the total number of terms in the sum minus one. For example, again referring to the last column of Table 2.1, the variance is

$$\frac{(21-10.21)^2+(4-10.21)^2+(3-10.21)^2+\cdots+(6-10.21)^2}{(24-1)}=48.87$$

This number is difficult to interpret, since it is on the scale of the square of days. Therefore, the square root of the variance, called the standard deviation, is often used. Here the standard deviation is equal to $\sqrt{48.87}=7.0$ days. Roughly speaking, the standard deviation can be considered as the average amount by which each observation can be expected to differ from the mean value, and so is a measure of dispersion in the population. Again, because of the missing value for patient #13, we divided by 23 rather than 24. You might wonder why we divide by n-1 rather than n, the total number of terms in the sum. The intuitive reason is that we would like to estimate the average dispersion from the true average, but the mean of 10.21 in the sum above is an estimate of the true average, and this estimate tends to lie slightly closer to the observed data points than they do to the true mean. Dividing by the smaller number 23 rather than 24 adjusts the variance to be a slightly larger value, to account for using the sample mean rather than the true mean.

Inter-Quartile Range

As mentioned in the above discussion relating to boxplots, the inter-quartile range provides upper and lower limits inside of which the middle 50% of the values in the sample lie.

Other descriptive statistics that are often useful are the minimum and maximum values in the sample, both for their description of the range found in the data, and also since they may be used to scan for errors in the data set by checking that all values are within a feasible range. While this does of course not guarantee that there have been no errors in compiling the data, gross errors may be found. Percentages are often used to summarize data denoting group membership (often called dichotomous data if there are two groups, and categorical data for more than two groups). For example, from Table 2.1 we can say that $^{13}/_{25}$ = 52% of the patients are in the laparoscopy group, and $^{24}/_{25}$ = 96% were male. With some variables, it may be useful to provide several different numerical summaries. While the average smoke years in Table 2.1 is 8.84 years, only $^{8}/_{25}$ = 32% of those sampled in fact smoke, and the average number of smoke years among those who smoke is in fact 27.625 years. The general rule with descriptive data is to use as many descriptive statistics as is necessary to provide an accurate summary of the data. Not doing so may present a false picture of the data, and is at the root of most misuses (both deliberate and nondeliberate!) of statistics.

3. Probability

In descriptive statistics, we are concerned with learning about the features of a particular data set or describing the observed relationship between variables. In inferential statistics, we would like to be able to draw conclusions about a population given data from a sample drawn from that population.

Consider Table 2.2, which shows ten different possible results from a hypothetical randomized clinical trial of 20 patients with pain following knee surgery. Half of the subjects were given Aspirin for their pain, while the rest were given Tylenol. The trials are listed in order of increasing evidence in favor of Aspirin. In Trial #1, it is easy to agree that there is no evidence to favor one drug over the other, while in trial 10, if the data are to be believed, Aspirin has clearly performed better than Tylenol. What, however, should one conclude following trials #2, #5, or #8? Inferential statistics attempts to numerically evaluate the evidence in any given trial, so that reasonable conclusions can be drawn.

In drawing conclusions, however, it is crucial to realize that the trial must be put into its' proper context. For example, suppose trial #10 in Table 2.2 was in fact observed. What should be concluded? While the data from these 20 subjects certainly favored Aspirin, there are many decades of evidence regarding Tylenol which must also be factored into any conclusions. In fact, I doubt that if such a trial were to be published tomorrow that Tylenol users around the world would immediately switch, nor would physicians stop recommending the use of this drug. Clearly the results of any current trial must be considered along with any past evidence in drawing final conclusions.

Along these lines, consider Figure 2.5, which illustrates an example due to Savage[2] which is also discussed by Berger.[3] The first illustration in Figure 2.5 shows a musicologist who claims to be able to correctly identify the works of Beethoven, as distinct from the works of Mozart, simply by looking at a page of their notated music. Suppose that you decide to put him to the test, and that in fact he is able to identify 5 out of 5 pieces of music correctly. In the bottom illustration, a psychic claims to be able to predict the outcome, heads or tails, of a flipped coin. Suppose she also gets 5 correct predictions in 5 tosses of a coin you take out of your pocket.

In the first instance, most observers would have no trouble believing the musicologist's claim. Many persons, however, would demand more evidence from the psychic, thinking that she may simply have been lucky. Note that the probability of guessing correctly on each trial is $^1/_2$ for both the musicologist and the psychic, so that the chances of getting five correct guesses in a row purely by chance is $(^1/_2)^5 = {}^1/_{32} = 0.03125$ or about 3%. Since each had the same chance of being correct by pure guesswork, it may seem "unobjective" or even unfair to demand more evidence from the psychic than we do from the musicologist, yet most of us would do so. The explanation is that we must draw conclusions based not only on the data at hand, but also by putting the current data into the larger context. Here the larger context includes our prior beliefs about the existence of music experts and psychics. In this

Table 2.2. Results from 10 hypothetical trials of Aspirin versus Tylenol

	Aspirin		Tylenol	
Trial	Cured	Not Cured	Cured	Not Cured
1	5	5	5	5
2	6	4	5	5
3	6	4	4	6
4	7	3	5	5
5	7	3	4	6
6	8	2	4	6
7	8	2	3	7
8	9	1	3	7
9	9	1	2	8
10	10	0	0	10

Fig. 2.5. The musicologist, the tea drinker, and the psychic.

light, it no longer seems surprising that we draw different conclusions, in that we know that many persons have studied to be music experts, while we may be skeptical that psychic powers exist. The middle illustration may represent a case intermediate to the other two in terms of prior beliefs, where a man claims to be able to tell whether the milk or water has been poured first into a cup of tea. If he is able to correctly identify 5 cups in a row as to whether the milk or water came first, some of us may still be skeptical (but perhaps not as skeptical as we were of the psychic's claims), and others may be more convinced (but perhaps not as convinced as we were of the musicologist's claim).

Similar situations arise in the interpretation of medical data, as might happen if a surprising result such as that of trial #10 occurred in Table 2.2, or in evaluating alternative therapies such as touch therapy.[4] In all trials, however, it is always important to consider not only the data at hand, but to also put the trial into the context of what is known from other sources. We will return to this point later in the Chapter, where we will see that the choice of whether to formally include information from outside the trial into the analysis or to only consider it informally in a post-hoc discussion has an impact on the type of statistical analysis that one will perform on any set of data.

3.1. What Is Probability?

The seemingly simple question of "what is probability?" has in fact been hotly debated by philosophers, statisticians and others for many decades with no general agreement. This question is important, since as we will see, it has direct implications for the type of statistical analysis to be performed, with different definitions leading to different schools of statistical inference. There are two main modes of statistical inference, usually referred to as frequentist or classical inference, and Bayesian inference. Many other ideas have also surfaced, for example pure likelihood methods[5] and belief functions,[6] although we will not discuss these further here.

The frequentist school defines the probability of an event as the number of times the event occurs divided by the number of trials, n, as n approaches infinity. For example, the probability that a coin will come up heads is 0.5, since assuming the coin is fair, as the number of trials (flips of the coin) gets larger and larger, the observed proportion will be, on average, closer and closer to 0.5. Similarly, the probability that a surgical technique is successful would be defined as the number of times it is observed to be successful in a large (theoretically infinite) number of trials.

While this definition has a certain logic to it, there are some problems. For example, what is the probability it will rain today? Since "today" is a unique event that will not happen an infinite number of times, the above definition cannot be applied. Nevertheless, we often hear statements such as "There is a 40% chance of rain today". Similarly, suppose that a new surgical technique has just been developed, and the surgeon is debating whether or not to apply it to his next patient. Surely the probability of success of the operation compared to the probability of success of the standard procedure for the patients condition will play a large role in the decision, but again, there are as yet no trials (and certainly not an infinite number of trials) upon which to define the probability. While we can conceptualize an infinite number of trials that may occur into the future, this does not help in defining a probability for today's decision as to which surgery to perform. Clearly, this definition is limited, not only in the case of events that can only happen once, but also because one rarely if ever can observe an infinity of like events.

The second school of thought, often referred to as the Bayesian school, defines the probability of any event occurring as the degree of belief that the event will occur. Therefore, based on the physics of the problem and perhaps a number of test flips, a Bayesian may assert that the probability of a coin flip coming up heads should be close to 0.5. Similarly, based on an assessment that may include both objective data and subjective beliefs about the surgical technique, the surgeon may assert that the probability that the surgery will be successful is 85%.

The obvious objection to Bayesian probability statements is that they are "subjective", so different surgeons may state different probabilities of success for the success rate of the surgery, and in general, there is no single "correct" probability statement that may be made about any event, since they reflect personal subjective beliefs. Supporters of the Bayesian viewpoint counter that the frequentist definition of probability is difficult to operationalize in practice, and does not apply to many important situations. Furthermore, the possible lack of agreement on the "correct" probability for any given event can be viewed as an advantage, since it will correctly mirror the range of beliefs that may exist for any event that does not have a large amount of data collected in which to accurately estimate its probability. Hence having a range of probabilities depending on the personal beliefs of a community of surgeons is a useful reflection of reality.

It may be surprising to many readers that there is no consensus in the statistical community about as basic a concept as a definition of probability, but such is the state of statistical thought as the 21st century begins. A majority of statistical analyses that appear in medical journals are performed using procedures that arise from the frequentist definition of probability, although the past 10 years have seen a rapid increase in Bayesian analyses in many applied fields, including medicine. Many statisticians use the full range of statistical procedures, including Bayesian and frequentist procedures, often switching between the two in analyzing the same data set. In this Chapter we will discuss both types of procedures, after first discussing some basic rules of probability that both schools of thought follow.

3.2. Rules of Probability

Whichever definition is used, probabilities should obey the following rules. Let Ω denote the set of all possible outcomes in a given experiment, and let E denote any event. The basic rules of probability are:

1. The probability of the set of all possible outcomes is $Pr\{\Omega\} = 1$.
2. All probabilities are between zero and one, so that $0 \le Pr\{E\} \le 1$ for every event E.
3. If events E and F are disjoint (that is, they have no outcomes in common), then the probability that either E or F occurs is given by $Pr\{E \cup F\} = Pr\{E\} + Pr\{F\}$, where $\cup$ denotes the union of the events E and F. Note that this rule implies that the probability of the complementary event, the event that E does not occur, usually denoted by E^c, must be equal to $1 - P\{E\}$.

4. If events E and F are independent (that is, knowing the outcome of E provides no information concerning the likelihood of F occurring), then the probability that events E and F both occur is given by $Pr\{E \cap F\} = Pr\{E\} \times Pr\{F\}$, where $\cap$ denotes the intersection of the events E and F.

Let us look at an example:

Example 1

Suppose that the probability of a certain type of surgery turning out successfully is 0.70 (that is, 70% chance of success).

1. What is the probability that the surgery will either be a success or a failure?
2. What is the failure rate for this operation?
3. If two persons undergo this surgery, what is the probability that both will have successful surgery?
4. What is the probability that exactly one of the two surgeries will be successful?

Solutions

1. Since the events of successful surgery and unsuccessful surgery together in fact make up the entire sample space (that is, one of these two events must happen, so that the union of these events must happen), by probability rule number 1, the probability of this event must be 1.
2. By rule number 3 for disjoint (in this case, complementary) events, the probability of a failure must be 1-0.7 = 0.3.
3. Since we can presume the events are independent, using rule 4 above we have 0.7 x 0.7 = 0.49. Thus there is a close to 50% chance that both surgeries will be successful.
4. There is a 70% chance that the first surgery will be successful, and a 30% chance that the second one will fail. Overall, then, there is a 0.7 x 0.3 = 0.21 chance of exactly one success in that order. In addition, however, there is a similar 0.21 chance that the first surgery will fail, and the second one will be a success. By rule 3, therefore, since these two events are disjoint, there is a 42% probability of exactly one success.

3.3. Probability Functions and Densities

Rather than working out all problems involving probabilities by first principles as above, short cut rules for common situations have been devised, leading to probability functions and probability densities. Probability functions are used for discrete variables (see Section 2.1), and simply provide the probability for each possible value that may occur. Probability densities are curves that cover the range of values for continuous variables. In any given region of the curve, the higher the curve in that region compared to others, the more likely it is that values in that range will occur. Technically, the area under the curve between any two points gives the probability of getting a value in that range. From probability rule 1, the total area under the curve over its' entire range must be equal to one. Since the area under any given single point is zero (since there is no width), the probability of any single point from a continuous density is zero. The term probability distribution is sometimes used as

2

a substitute for probability functions or probability densities, although somewhat confusingly, the term is also used for the cumulative distribution for both discrete and continuous variables. Cumulative distributions provide the probability of obtaining a result equal to or less than a given value of that variable.

We will now look at two examples of commonly used probability functions and probability densities. The first of these, the binomial distribution, is a discrete probability function, while the second, the normal distribution, is continuous. One further continuous probability density, the beta density, is introduced in Section 9 below. We briefly mention the use of the χ^2 and t-densities in Sections 5 and 6, respectively.

3.4. The Binomial Distribution

One of the most commonly used probability functions is the binomial. The binomial probability function allows one to calculate the probability of obtaining a given number of "successes" in a given number of independent trials. In general, the formula for the binomial probability function is:

$$Pr\{x \text{ "successes" in } n \text{ "trials"}\} = \frac{n!}{x!(n-x)!} p^x (1-p)^{n-x} \quad (1)$$

where $n!$ is read "n factorial", and is shorthand for

$$n \times (n-1) \times (n-2) \times (n-3) \times \cdots \times 3 \times 2 \times 1.$$

For example, $5! = 5 \times 4 \times 3 \times 2 \times 1 = 120$, and so on. By convention, $0! = 1$. The probability of a success on each trial is assumed to be p. Suppose we wish to calculate the probability of $x = 8$ successful surgeries in $n = 10$ operations, where the probability of a successful operation each time is 70%. From the binomial formula, we can calculate:

$$\frac{10!}{8!2!} 0.7^8 (1-0.7)^2 = 0.2335$$

so that there is a little bit less than a one in four chance of getting 8 successful surgeries in 10 trials. Similarly, the probability of getting 8 or more (that is, 8 or 9 or 10) successful surgeries is found by adding up three probabilities of the above type. As an exercise, you can check that this probability is 0.3829.

The binomial distribution has a theoretical mean of $n \times p$, which in a nice intuitive result. For example, if you perform $n = 100$ trials, and on each trial the probability of success is, say, $p = 0.4$ or 40%, then you would intuitively expect $100 \times 0.4 = 40$ successes. The variance of a binomial distribution is $n \times p \times (1-p)$, so that in the above example it would be $100 \times 0.4 \times 0.6 = 24$. Thus the standard deviation is $\sqrt{24} = 4.90$, roughly meaning that while on average one expects about 40 successes, one also expects each result to deviate from 40 by an average of about 5 successes.

3.5. The Normal Distribution

Perhaps the most common distribution used in statistical practice is the normal distribution. The normal distribution is the familiar "bell-shaped" curve, as seen in Figure 2.6. Technically, the curve is traced out by the normal density function:

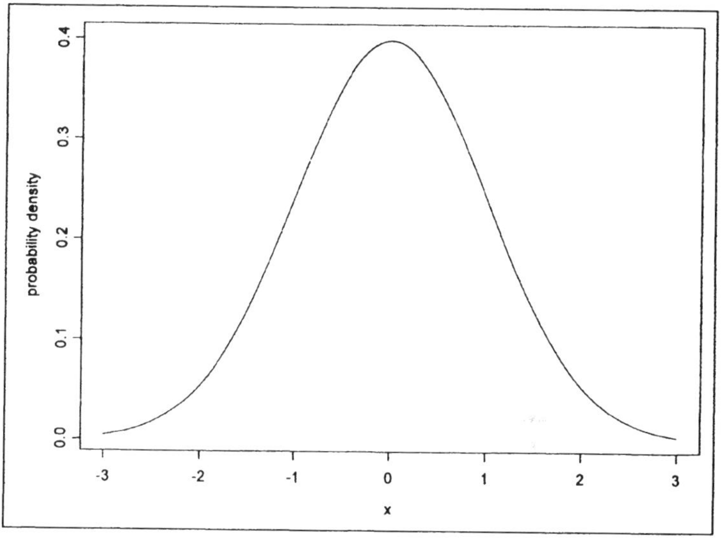

Fig. 2.6. The standard normal distribution with mean $\mu = 0$ and standard deviation sigma = 1. Approximately 95% of the area under the curve falls within 2 standard deviations on either side of the mean, and about 68% of the area falls within one standard deviation from the mean.

$$\frac{1}{\sqrt{2\pi}\sigma}\exp\left\{-\frac{1}{2}\frac{(x-\mu)^2}{\sigma^2}\right\} \tag{2}$$

where "exp" denotes the exponential function to the base $e = 2.71828$. The Greek letter μ is the mean of the normal distribution (set to zero in the standard normal curve of Fig. 2.6), and the standard deviation is σ, set to 1 in the standard normal curve. While Figure 2.6 lists the standard version of the normal curve ($\mu = 0$, $\sigma^2 = \sigma = 1$), more generally, the mean μ can be any real number and the standard deviation can be any number greater than 0. Changing the mean shifts the curve depicted in Figure 2.6 to the left or right so that it remains centered at the mean, while changing the standard deviation stretches or shrinks the curve around the mean, all while keeping its' bell shape. Note that the mean, median and mode (most likely value, i.e., highest point on the curve) of a normal distribution are always the same and equal to μ.

The normal density function has been used to represent the distribution of many measures in medicine. For example, blood pressures, cholesterol levels or bone mineral densities in a given population may be said to follow a normal distribution with a given mean and standard deviation. It is very unlikely that any of these or other quantities exactly follow a normal distribution. For instance, none of the above-mentioned quantities can have negative numbers, while the range of the normal distribution always includes all negative (and all positive) numbers. Nevertheless,

2

Table 2.3. Table of standard normal distribution probabilities. Each number in the table provides the probability that a standard normal random variable will be less than the number indicated.

	0.00	0.01	0.02	0.03	0.04	0.05	0.06	0.07	0.08	0.09
0.0	0.5000	0.5040	0.5080	0.5120	0.5160	0.5199	0.5239	0.5279	0.5319	0.5359
0.1	0.5398	0.5438	0.5478	0.5517	0.5557	0.5596	0.5636	0.5675	0.5714	0.5753
0.2	0.5793	0.5832	0.5871	0.5910	0.5948	0.5987	0.6026	0.6064	0.6103	0.6141
0.3	0.6179	0.6217	0.6255	0.6293	0.6331	0.6368	0.6406	0.6443	0.6480	0.6517
0.4	0.6554	0.6591	0.6628	0.6664	0.6700	0.6736	0.6772	0.6808	0.6844	0.6879
0.5	0.6915	0.6950	0.6985	0.7019	0.7054	0.7088	0.7123	0.7157	0.7190	0.7224
0.6	0.7257	0.7291	0.7324	0.7357	0.7389	0.7422	0.7454	0.7486	0.7517	0.7549
0.7	0.7580	0.7611	0.7642	0.7673	0.7704	0.7734	0.7764	0.7794	0.7823	0.7852
0.8	0.7881	0.7910	0.7939	0.7967	0.7995	0.8023	0.8051	0.8078	0.8106	0.8133
0.9	0.8159	0.8186	0.8212	0.8238	0.8264	0.8289	0.8315	0.8340	0.8365	0.8389
1.0	0.8413	0.8438	0.8461	0.8485	0.8508	0.8531	0.8554	0.8577	0.8599	0.8621
1.1	0.8643	0.8665	0.8686	0.8708	0.8729	0.8749	0.8770	0.8790	0.8810	0.8830
1.2	0.8849	0.8869	0.8888	0.8907	0.8925	0.8944	0.8962	0.8980	0.8997	0.9015
1.3	0.9032	0.9049	0.9066	0.9082	0.9099	0.9115	0.9131	0.9147	0.9162	0.9177

1.4	0.9192	0.9207	0.9222	0.9236	0.9251	0.9265	0.9279	0.9292	0.9306	0.9319
	0.00	0.01	0.02	0.03	0.04	0.05	0.06	0.07	0.08	0.09
1.5	0.9332	0.9345	0.9357	0.9370	0.9382	0.9394	0.9406	0.9418	0.9429	0.9441
1.6	0.9452	0.9463	0.9474	0.9484	0.9495	0.9505	0.9515	0.9525	0.9535	0.9545
1.7	0.9554	0.9564	0.9573	0.9582	0.9591	0.9599	0.9608	0.9616	0.9625	0.9633
1.8	0.9641	0.9649	0.9656	0.9664	0.9671	0.9678	0.9686	0.9693	0.9699	0.9706
1.9	0.9713	0.9719	0.9726	0.9732	0.9738	0.9744	0.9750	0.9756	0.9761	0.9767
2.0	0.9772	0.9778	0.9783	0.9788	0.9793	0.9798	0.9803	0.9808	0.9812	0.9817
2.1	0.9821	0.9826	0.9830	0.9834	0.9838	0.9842	0.9846	0.9850	0.9854	0.9857
2.2	0.9861	0.9864	0.9868	0.9871	0.9875	0.9878	0.9881	0.9884	0.9887	0.9890
2.3	0.9893	0.9896	0.9898	0.9901	0.9904	0.9906	0.9909	0.9911	0.9913	0.9916
2.4	0.9918	0.9920	0.9922	0.9925	0.9927	0.9929	0.9931	0.9932	0.9934	0.9936
2.5	0.9938	0.9940	0.9941	0.9943	0.9945	0.9946	0.9948	0.9949	0.9951	0.9952
2.6	0.9953	0.9955	0.9956	0.9957	0.9959	0.9960	0.9961	0.9962	0.9963	0.9964
2.7	0.9965	0.9966	0.9967	0.9968	0.9969	0.9970	0.9971	0.9972	0.9973	0.9974
2.8	0.9974	0.9975	0.9976	0.9977	0.9977	0.9978	0.9979	0.9979	0.9980	0.9981
2.9	0.9981	0.9982	0.9982	0.9983	0.9984	0.9984	0.9985	0.9985	0.9986	0.9986
3.0	0.9987	0.9987	0.9987	0.9988	0.9988	0.9989	0.9989	0.9989	0.9990	0.9990

for appropriately chosen mean and standard deviation, the probability of out of range numbers will be vanishingly small, so that this may be of little concern in practice. We may say, for example, that diastolic blood pressure in a given population follows a normal distribution with mean of 80 and a standard deviation of 10, so that the probability of a value less than zero is 6.2×10^{-16}, and in fact the probability of being less than 50 (three standard deviations below the mean) is only 0.0013. In the words of statistician George Box, "All models are wrong, but some are useful."[7]

To calculate probabilities associated with the normal distribution, one must find the area under the normal curve. Since this is mathematically difficult, normal tables are usually used, as presented in Table 2.3. To see how Table 2.3 is used, consider the following example:

Example 2

What is the probability that a standard normal random variable falls in between -.3 and +1.23?

Solution

Table 2.3 presents the values for the area under standard normal curve from $-\infty$ up to the given number in the table (in other words, they present the cumulative distribution function, see Section 2.3). Since the normal curve is symmetric about 0, the area under the curve above +0.3 is the same as the area under the curve below -0.3. Therefore, normal tables need only present areas for positive numbers. To calculate the area we need, we first note that if we could find the area to the left of +1.23 and subtract from it the area to the left of -0.3, then we will be left with the area we need (you may wish to draw a little sketch to convince yourself of this). Now: the area to the left of +1.23 is available directly from Table 2.3, it is 0.8907. Next, remembering the probability rule for complementary events, the area to the right of +0.3 must be one minus the area to the left of +0.3. Looking this up on the table gives 0.3821 (= 1-0.6179), which must also be the area to the left of -0.3. Therefore, the probability that we are looking for in this problem must be 0.8907-0.3821 = 0.5086.

Example 3

In a normal curve with $\mu = 80$ and $\sigma = 10$, what is the probability of falling above 95?

Solution

We are no longer in the situation where we can use Table 2.3 directly, since we do not have a standard normal curve. However, since all normal curves have the same basic shape, we can map our problem into one involving the standard normal curve by standardizing. Standardizing involves transforming the value 95 to where its relative position would be on the standard normal curve. For example, we can see that 95 is exactly 1.5 standard deviations above the mean of the normal curve whose mean is 80 and whose standard deviation is 10. Therefore, we can look up the area below 1.5 on Table 2.3, which is 0.9332, so that the area above 1.5 is 1-0.9332 = 0.0668, which is our desired probability. In general, one standardizes by subtracting the mean and then dividing by the standard deviation. Following this rule we would have

$$z = \frac{95 - 80}{10} = 1.5$$

where by convention z represents the standardized value.

3.6. Normal Approximation to the Binomial Distribution

Recall that the formula for the binomial distribution is given by

$$Pr\{x \text{ success in } n \text{ trials}\} = \frac{n!}{(n-x)!x!} p^x (1-p)^{(n-x)}$$

We can either calculate this directly, by plugging numbers in the above formula, or look up the result in tables of the binomial distribution, available in many statistical textbooks. What happens, however, if we wish to know the probability of getting $X = 80$ or more successes in $N=150$ trials, with $\pi = 0.6$? Binomial tables do not generally go that high, and calculations seem infeasible, as, for example, 150! is a 263 digit number, and 0.6^{80} is a very small number, and most calculators/computer programs do not handle these extreme numbers very well. In addition, one would have to sum 71 of these numbers to get the final answer.

The solution is to approximate the required binomial probabilities by a normal distribution, and then look up the probabilities using tables of normal probabilities. This approximation works well as long as the sample size is large. An often quoted rule of thumb is that both $n \times p$ and $n \times (1-p)$ need to be greater than 5. We proceed as follows:

1. Find the mean and variance of the binomial distribution of interest. In the above example,

   $$\mu = n \times p = 150 \times 0.6 = 90$$

   and

   $$\sigma^2 = n \times p \times (1 - p) = 150 \times 0.6 \times 0.4 = 36.$$

2. Then reason as follows. The binomial distribution we have is close to a normal distribution with $\mu = 90$ and $\sigma^2 = 36$. Therefore, the binomial probability we want is close to the normal probability that $X \geq 80$. Standardizing by taking

   $$z = \frac{80 - 90}{\sqrt{36}} = -1.67,$$

   and looking up the result in Table 2.3 gives 0.9525.

The value 80 is sometimes changed to 79.5, called the continuity correction. It is used to make the approximation slightly more accurate. With the continuity correction, the probability would have been 0.9599, while the exact answer is in fact 0.9591.

To summarize, when the sample size is large, binomial probabilities are well approximated by areas under the appropriate normal curve. We therefore convert the binomial problem to one involving the area under a normal curve, using the normal curve with the same mean and variance as our original binomial distribution.

3.7. Central Limit Theorem

The binomial distribution is not the only case where a normal distribution could be used to approximate another distribution when the sample size grows large. Consider taking a random sample of 500 patients visiting their family physician for their periodic health exam. If the blood pressure of each patient is recorded and an average is taken, one could use this value as an estimate of the average in the population of all patients who might visit their family physician for a routine checkup. However, if this experiment was repeated a second time, it would be quite unexpected for the second average of 500 subjects to be identical to the first average, although one could expect it to be close.

How these averages vary from one sample to another is given by the Central Limit Theorem, which in its simplest form states:

Central Limit Theorem

Suppose that a population has true (but possibly unknown) mean μ and standard deviation σ. The distribution of the sample average, $\bar{x}$, approaches a normal distribution as the sample size grows large, with mean μ and standard deviation $\frac{\sigma}{\sqrt{n}}$. The standard deviation about a sample mean, $\frac{\sigma}{\sqrt{n}}$, is often called the standard error.

This useful theorem has two immediate consequences. First, it accounts for the popularity of the normal distribution in statistical practice. Even if an underlying distribution in a population is non-normal (for example, if it is skewed), the distribution of the sample average from this population becomes close to normal, if the sample size is large enough. Second, the result connects the sample mean to the population mean, forming the basis for much of statistical inference. In particular, notice that as the sample size n increases, the standard deviation (standard error) $\frac{\sigma}{\sqrt{n}}$ of the sample mean around the true mean, decreases, so that on average the sample mean $\bar{x}$ gets closer and closer to μ.

This ends our brief tour of the world of probability. Armed with these probability basics, we are now ready to consider some simple statistical inferences.

4. Drawing Statistical Inferences About a Population from a Sample

In this section we will begin to consider how to draw inferences about populations by statistically analyzing samples of data using standard frequentist methods. We will learn how to correctly interpret *p*-values and confidence intervals, in the context of a simple example. Sections 5 and 6 which follow will apply these ideas to problems with proportions and means, respectively. Section 9 will introduce Bayesian techniques for statistical inference.

4.1. The Types of Questions that Can Be Answered by Standard Statistical Inferences

Consider the following situation: A surgeon knows from past experience that a certain standard type of surgery has a success rate of 80%. He has just read in the literature, however, that a new promising type of surgery was tested on a group of 10 patients, and had a cure rate of 90% (9 out of the 10 patients were cured). Should he switch to the new type of surgery for his next patient?

Surely, assuming he knows how to perform the new technique, the surgeon should switch if the true cure rate for the new surgery is better than the known standard rate of 80%. The observed rate has so far been 90%, but the surgeon knows that this is based on only a small sample size. How sure is he that the rate of the new surgery will be over 80% based on the limited data collected so far?

Consider the following questions that the surgeon may (or may not!) find relevant to his decision:

1. What is the probability that the new treatment has a rate above 80%, given the data just observed and any other relevant knowledge?
2. Can an interval be found such that there is a 95% probability that the true success rate lies inside this interval, given the data just observed and any other relevant knowledge?
3. Suppose for the moment that the success rate of the new surgical treatment is 80% exactly. This represents a null hypothesis where the new surgical treatment has exactly the same rate as the standard surgical treatment. Under these conditions, what is the probability of observing 9 or more successes out of the 10 subjects? In other words, what is the probability of what we observed (9 out of 10) and all outcomes more extreme than what was observed (in this case, only one more extreme outcome, 10 out of 10).
4. Can a procedure be found such that intervals constructed via the procedure will contain the true proportion of interest 95% of the time the procedure is applied in various different problems?

Take some time to ponder which of these questions you think the surgeon would most want answered in making a decision about which treatment to use. Most people would find the first two questions rather natural. Question 1 is perhaps the most direct statement of what is desired, but question 2 could provide useful information about whether the rate is more likely to be near 81% or 98%, say, which could influence decision making if the new treatment is more costly. Questions 3 and 4 are at least somewhat obscure and at best of indirect interest. Why would one care to know the probabilities of more extreme data that in fact did not occur? And why fix the rates as exactly equal, when almost surely they are not exactly equivalent? Why the strange indirect formulation of question 4, which refers to problems other than the one of current interest, compared to the direct simplicity of question 2?

You may then be surprised to know that if you looked into the surgical and most other medical literature today, you would almost certainly find the answer to question number 3, possibly the answer to question number 4, but almost never the answers to questions 1 or 2! Question 3 is addressed by providing a *p*-value via a frequentist hypothesis test, while providing a confidence interval addresses question 4. These procedures will be discussed in Sections 4.2 and 4.3, respectively. Questions 1 and 2 are associated with Bayesian inference, which will be covered in Section 9.

4.2. Standard Frequentist Hypothesis Testing

Suppose we wish to test the null hypothesis that the new surgical technique is in fact not better than the standard technique, versus an alternative hypothesis that it is better. Formally, we can state these hypotheses as:

$$H_0: p \leq 0.8$$
$$H_A: p > 0.8$$

where p represents the unknown true probability of success of the new surgical treatment. In considering hypothesis testing, we note that there are four possible results, as shown in the table below.

		True State of Nature	
		H_A, +	H_0, –
Test	+ (Reject H_0)	1-β	α
	– (Do not Reject H_0)	β	1-α

According to the table, if the new surgical technique is in fact better than the standard and we reject the null hypothesis, then we have made a correct decision, as also happens if the null hypothesis is in fact correct, and we do not reject it. On the other hand, if we reject the null hypothesis as false when it is in fact true, we make a so-called Type-I error, also sometimes called an α error, and if we fail to reject the null hypothesis when it is in fact false, we make a Type-II, or β error. More precisely, α is the probability of making a Type I error when the null hypothesis is true, and β is the probability of making a Type II error when the alternative hypothesis is true. The power of a study is defined as the probability of rejecting the null hypothesis when the alternative hypothesis is in fact true, so that the power is equal to 1-β. To summarize, we have:

$$\alpha = Pr\{\text{rejecting } H_0 | H_0 \text{ is true}\} = \text{type I error}$$
$$1\text{-}\alpha = Pr\{\text{not rejecting } H_0 | H_0 \text{ is true}\}$$
$$\beta = Pr\{\text{not rejecting } H_0 | H_A \text{ is true}\} = \text{type II error}$$
$$1 - \beta = Pr\{\text{rejecting } H_0 | H_A \text{ is true}\} = \text{Power}$$

Probabilities written in the form of Pr{A|B} are called "conditional probabilities", and the notation is read as the probability that the event A occurs, given that the event B is known to have occurred. Thus all of the quantities above are conditional on knowing whether the null or alternative hypotheses are in fact true. Of course, we generally do not know whether the null hypothesis is true or not, so that these conditional statements are at best of indirect interest. Once we obtain our data, we would ideally like to know the probability that the null hypothesis is true, not assume the null hypothesis is true! See Section 9 for further discussion of this point.

Although it is important to understand the types of errors that can be made when hypothesis testing, the result of a hypothesis test is usually reported as a p-value, which we now define.

Definition

The p-value is the probability of obtaining a result as or more extreme than that observed assuming that the null hypothesis is in fact true.

It is very important to note that the p-value is not the probability that the null hypothesis is correct after having seen the data, even though many clinicians often falsely interpret it this way. The p-value does not directly or indirectly provide this probability, and in fact can be orders of magnitude different from it. In other words, it is possible to have a p-value equal to 0.05, when the probability of the null hypothesis is 0.5, different from the p-value by a factor of 10. Therefore, p-values are the answer to the rather obscure question number 3 above, which, at best, indirectly helps the surgeon in the decision as to which technique to use.

Given the correct definition of a p-value, how would we calculate it? For our example of the new surgical technique, the definition implies that we need to calculate the probability of obtaining 9 or 10 successful surgeries in the 10 patients to whom it was applied, given that the true rate of success is exactly 80%. From the binomial distribution (see equation (1)), this can be calculated as

$$\begin{aligned} p &= \frac{10!}{9!1!}0.8^9(1-0.8)^1 + \frac{10!}{10!0!}0.8^{10}(1-0.8)^0 \\ &= 0.2684 + 0.1074 \\ &= 0.3758. \end{aligned}$$

So there is about a 37.6% chance of obtaining results as or more extreme than the 9 out of 10 result observed, if the true rate for the new technique is exactly 80%. Therefore, the observed result is not unusual, and hence compatible with the null hypothesis, so that we cannot reject H_0. Notice that if we had observed the same success rate but with a larger sample size of 100, the p-value would have been 0.006. This is calculated by finding the probability of 90 or more successes in 100 trials, and using the normal approximation to the binomial distribution given in Section 3.7. With a sample size of 100, the event of the observed data or data more extreme would be a rare event if the null hypothesis were true, so that the null hypothesis could be rejected. Therefore, p-values depend not only on the observed success rate, but also on the sample size.

While p-values are still often found in the literature, there are several major problems associated with their use:

1. As mentioned above, they are often misinterpreted as the probability of the null hypothesis given the data, when in fact they are calculated assuming the null hypothesis to be true.
2. Clinicians often use them to "dichotomize" results into "important" or "unimportant" depending on whether $p < 0.05$ or $p > 0.05$, respectively. However, there is not much difference between p-values of 0.049 and 0.051, so that the cutoff of 0.05 is arbitrary.
3. P-values concentrate attention away from the magnitude of treatment differences. For example, one could have a p-value that is very small, but is associated with a clinically unimportant difference. This is especially prone to occur in cases where the sample size is large. Conversely, results of potentially great clinical interest are not necessarily ruled out if $p > 0.05$, especially in studies with small sample sizes. Therefore, one should not confuse statistical significance (i.e., $p < 0.05$) with practical or clinical importance.

2

2

4. The null hypothesis is almost never exactly true. In the above example, does one seriously think that the new success rate could be exactly 80% (rather than, say, 80.0001% or 79.9999%)? Since one knows the null hypothesis is almost surely false to begin with, it makes little sense to test it. Instead, one should concern oneself with the question of "by how much are the two treatments different".

There are so many problems associated with p-values that most statisticians now recommend against their use, in favor of confidence intervals or Bayesian methods. In fact, some prominent journals no longer publish p-values at all,[8] others strongly discourage their use[9] and many others have published articles and editorials encouraging the use of Bayesian methodology.[10,11] We will cover these more informative techniques for drawing statistical inferences, starting with confidence intervals.

4.3. Frequentist Confidence Intervals

While the p-value provides some information concerning the rarity of events as or more extreme than that observed assuming the null hypothesis to be exactly true, it provides no information about what the true rate might be. In our example, we have observed a rate of 90%, but know that this is based on a small sample size, and that the observed rate may well be seen to increase or decrease as more data accumulates. Based on this data, however, what can we say about where we would expect the true rate to be?

One way to answer this question is with a confidence interval. Confidence intervals usually have the form

$$\text{estimate} \pm k \text{ x standard error}$$

where the estimate and standard error are calculated from the data, and where k is a constant with a value usually near 2.

For example, suppose one observes $x = 80$ successful surgical procedures in $n = 100$ operations, leading to an estimate of the success rate of $\hat{p} = \frac{x}{n} = 0.8$. We use the notation $\hat{p}$ rather than p to indicate that this is an estimated rate, not necessarily equal to the true rate, which we denote by p. Following the general formula above, a confidence interval for a binomial probability of success parameter is given by

$$\left(\hat{p} - z \times \sqrt{\frac{\hat{p} \times (1-\hat{p})}{n}}, \hat{p} + z \times \sqrt{\frac{\hat{p} \times (1-\hat{p})}{n}}\right) \quad (3)$$

where z is derived from Normal Tables as in Table 2.3, and is given by $z = 1.96$ for the usual 95% confidence interval. Therefore, the 95% confidence interval in our example is:

$$\left(0.8 - 1.96 \times \sqrt{\frac{0.8 \times 0.2}{100}}, 0.8 + 1.96 \times \sqrt{\frac{0.8 \times 0.2}{100}}\right)$$

which here gives (0.72, 0.88).

How does one interpret this confidence interval? Equation (3) provides a procedure that when used repeatedly across different problems, will capture the true value of p 95% of the time, and fail to capture the true value 5% of the time. In this sense, we have confidence that the procedure works well in the long run, although in any single application, of course, the interval either does or does not contain the true proportion p. Note that we are careful not to say that our confidence interval has a 95% probability of containing the true parameter value. In other words, we did not say that the true proportion of successful surgeries is in the interval (0.72, 0.88) with 95% probability. This is because the confidence limits and the true rate are both fixed numbers, and it makes no more sense to say that the true rate is in this interval than it does to say that the number 2 is inside the interval (1,6) with probability 95%. Of course, 2 is inside this interval, just like the number 8 is outside of the interval (1,6). However, in the procedure that we used to calculate the above confidence interval, we derived random upper and lower limits (as given by the formula in equation (3)), and in repeated uses of this formula across a range of problems, we expect the random limits to capture the true value 95% of the time, and exclude the true limit 5% of the time. Refer to Figure 2.7. If we look at the set of confidence intervals as a whole, we see that about 95% of them include the true parameter value. However, if we pick out a single trial, it either contains the true value (about 95% of the time) or excludes this value (about 5% of the time).

Despite their somewhat unnatural interpretation, confidence intervals are generally preferred to p-values. This is because they focus attention on the range of values compatible with the data, on a scale of direct clinical interest. Given a confidence interval, one can assess the clinical meaningfulness of the result, as can be seen in Figure 2.8.

Depending on where the upper and lower confidence interval limits fall in relation to the upper and lower limits of the region of clinical equivalence, different conclusions should be drawn. The region of clinical equivalence, sometimes called the region of clinical indifference, is the region inside of which two treatments, say, would be considered to be the same for all practical purposes. The point 0, indicating no difference in results between two treatments, is usually included in the region of clinical equivalence, but values above and below 0 are usually also included. How wide this region is depends on each individual clinical situation. For example, if one treatment is much more expensive than another, one may want at least a 5% advantage in order to consider it the preferred treatment (see Chapter 9). From Figure 2.8, there are five different conclusions that can be made after a confidence interval has been calculated:

1. The CI includes zero, and both upper and lower CI limits, if they were the true values, would not be clinically interesting. Therefore, this variable has been shown to have no important effect.
2. The CI includes zero, but one or both of the upper or lower CI limits, if they were the true values, would be interesting clinically. Therefore, the results of this variable in this study is inconclusive, and further evidence needs to be collected.
3. The CI does not include zero, and all values inside the upper and lower CI limits, if they were the true values, would be clinically interesting.

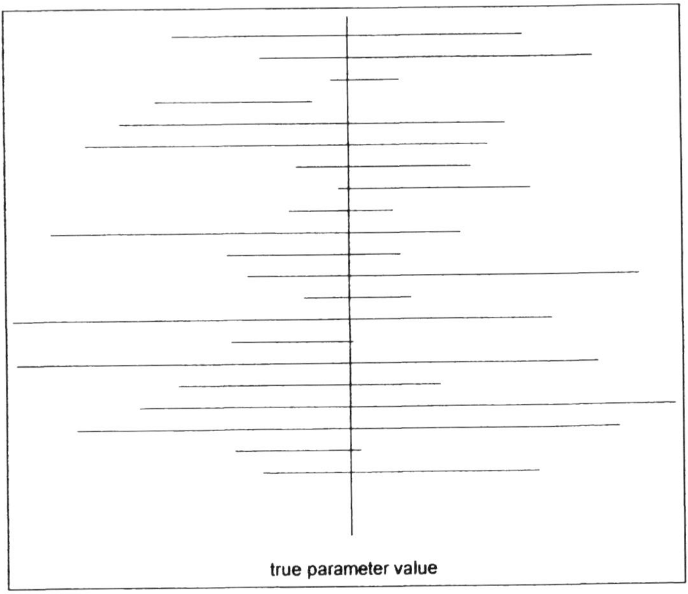

Fig. 2.7. A series of 95% confidence intervals for an unknown parameter.

Therefore, this study shows this variable to be important.

4. The CI does not include zero, but all values inside the upper and lower CI limits, if they were the true values, would not be clinically interesting. Therefore, this study shows this variable, while having some small effect, is not clinically important.
5. The CI does not include zero, but only some of the values inside the upper and lower CI limits, if they were the true values, would be clinically interesting. Therefore, this study shows this variable has at least a small effect, and may be clinically important. Further study is required in order to better estimate the magnitude of this effect.

For our problem, with 9 successful surgeries in 10 trials, the 95% confidence interval ranges from 55.5% to 99.7%, providing a large and inconclusive interval. [Technical note: Since the sample size is only 10 here, we used an "exact" formula different from that given by equation (3), which works better than (3) for small sample sizes.] More information would need to be provided in order to determine whether the new surgical technique is better or worse than the standard 80% success rate. Had the same rate been observed in 100 trials, for example, the confidence interval would have been (82.6%, 94.5%). Since this latter confidence interval excludes the null value of 80%, we know that the new technique is better than the old technique. Whether it is better enough to switch or not depends on whether we have a confidence interval of type 3 or type 5 (presumably we are not in a type 4 situation since an improvement in the success rate of almost 15% would be important). This is a clinical judgement that depends on many factors, including the costs

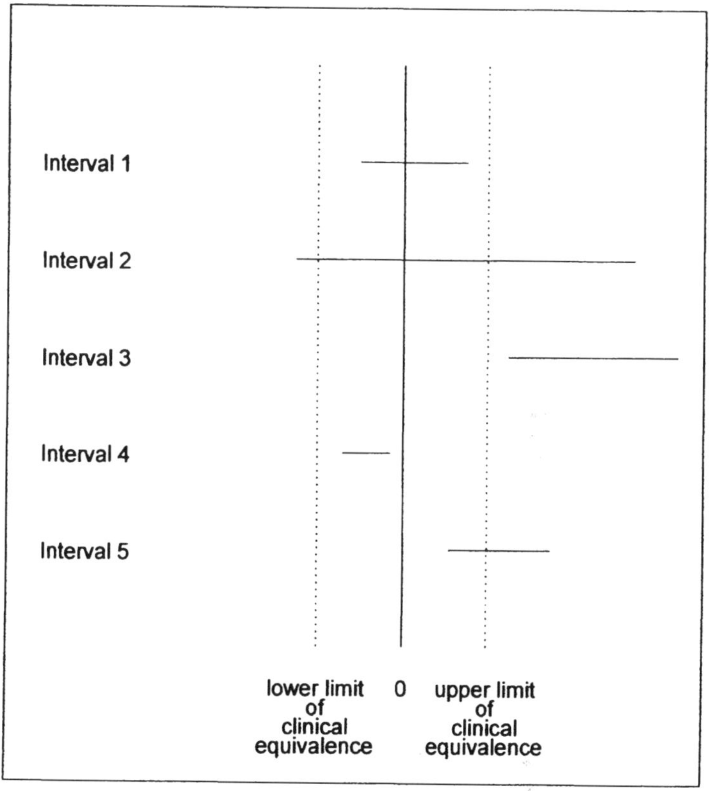

Fig. 2.8. How to interpret confidence intervals. Depending on where the confidence interval lies in relation to a region of clinical equivalence, different conclusions can be drawn.

and availabilities of the treatments, and possible undesirable side-effects of the new technique. The complexity of these factors makes finding a definitive region of clinical equivalence difficult in most situations, so that the rigorous application of the above guidelines for interpreting confidence intervals is rarely possible. Nevertheless, it is crucial to relate the location of any confidence interval to a region of clinical equivalence, whether explicit or implicit, in interpreting results from any medical trial or experiment.

4.4. Summary of Frequentist Statistical Inference

The main tools for statistical inference from the frequentist point of view are *p*-values and confidence intervals. *P*-values have fallen out of favor among statisticians, and although they continue to appear in a large proportion of medical journal articles, their use is likely to greatly diminish in the coming years. Confidence intervals provide much more clinically useful information than *p*-values, so are to be preferred in

2

practice. Confidence intervals still do not allow for the formal incorporation of pre-existing knowledge into any final conclusions. For example, in some cases there may be compelling medical reasons why a new technique may be better than a standard technique, so that faced with an inconclusive confidence interval, a surgeon may still wish to switch to the new technique, at least until more data become available. On what basis could this decision be justified? We will return to this question in Section 9, where we look at Bayesian statistical inference.

While we have so far discussed *p*-values and confidence intervals in the situation where data about a single success rate or proportion was of interest, similar techniques are available when comparisons of two or more proportions or inference about one or more means are of interest. While one Chapter cannot cover all such techniques in any detail, in the next two sections we will briefly present some tests that could be applied and confidence intervals that could be calculated for the most commonly occurring situations. In all cases, the interpretations of *p*-values and confidence intervals remain as discussed above.

5. Statistical Inference for Two or More Proportions

Consider the following example, adapted from Garraway et al[12] concerning the use of a specialized Stroke Unit versus a Medical Unit following an acute stroke in the elderly:

	Patient Independent	Patient Dependent	Total
Stroke Unit	67	34	101
Medical Unit	46	45	91
Total	113	79	192

One would like to draw inferences about whether a specialized stroke unit increases the proportion of independent patients following an acute stroke compared to the usual care of a Medical Unit. While one observes $\hat{p}_1 = 0.67$ probability of success in the Stroke Unit compared to a $\hat{p}_2 = 0.51$ rate in the Medical Unit for a 16% observed difference, we know from Section 4 that a confidence interval will provide us with a range of values that will help draw a better conclusion compared to simply looking at the observed difference. To calculate a confidence interval for this difference in proportions, we can use the following formula, which extends equation (3) to the case of two proportions:

$$\left(\hat{p}_1 - \hat{p}_2 - z^* \sqrt{\frac{\hat{p}_1(1-\hat{p}_1)}{n_1} + \frac{\hat{p}_2(1-\hat{p}_2)}{n_2}},\ \hat{p}_1 - \hat{p}_2 + z^* \sqrt{\frac{\hat{p}_1(1-\hat{p}_1)}{n_1} + \frac{\hat{p}_2(1-\hat{p}_2)}{n_2}}\right) \quad (4)$$

where $\hat{p}_1$ and $\hat{p}_2$ are the observed proportions in the two groups out of sample sizes n_1 and n_2, respectively, and z^* is the relevant percentile from normal tables, chosen according to the desired level of the confidence interval. For example, for a 95% confidence interval $z^* = 1.96$, for a 90% interval $z^* = 1.64$, and so on. Using the above formula for the stroke data given above, one finds that a 95% confidence interval for the difference in rates is (0.02, 0.30). This interval suggests that the Stroke Unit is indeed better, but if the true difference is in fact as low as 2%, we may decide it is not worth the extra expense, while if the true rates differ by 30%, we surely would find the Stroke Unit worthwhile. Unless cost is not a factor, further research may be required to narrow the confidence interval so that a more definitive decision could be reached.

Although confidence intervals are preferred for reasons discussed in Section 4, we will also discuss hypothesis testing for proportions, since one often sees such tests in the literature. Suppose we wish to test the null hypothesis that $p_1 = p_2$, that is, the null hypothesis states that the success rates are identical in the two units. Since we hypothesize $p_1 = p_2$, we expect to observe the following table of data, on average:

	Patient Independent	Patient Dependent	Total
Stroke Unit	59.44	41.56	101
Medical Unit	53.56	37.44	91
Total	113	79	192

Why do we "expect" to observe this table of data if the null hypothesis is true? We have observed a total number of 113 "successes" (independent patients) divided among the two groups. If $p_1 = p_2$ and if the sample sizes were equal in the two groups, we would have expected $^{113}/_{2} = 56.5$ successes in each group. However, since the sample sizes are not equal, we expect $113 \times {}^{101}/_{192} = 59.44$ to go to the Stroke Unit group, and $113 \times {}^{91}/_{192} = 53.56$ to go the Medical Unit group. Similarly, expected values for the dependent patients can be calculated. Observed discrepancies from these expected values are evidence against the null hypothesis. To perform a χ^2 (read "chi-squared") test, we now calculate:

$$X^2 = \sum_{\text{all cells}} \frac{(\text{observed} - \text{expected})^2}{\text{expected}}$$

$$= \frac{(67-59.44)^2}{59.44} + \frac{(34-41.56)^2}{41.56} + \frac{(46-53.56)^2}{53.56} + \frac{(45-37.44)^2}{37.44}$$

$$= 4.927$$

Comparing the $X^2 = 4.927$ value on χ^2 tables with 1 df (see Table 2.4), we find that $0.025 < p < 0.05$ (because our observed value, 4.927, lies between the 3.841 and 5.023 values in the table), so that we have evidence to reject the null hypothesis. This coincides with our conclusion from the confidence interval, but note that the confidence interval is more informative than simply looking at the p-value from the χ^2 test, since a range for the difference in rates is provided by the confidence interval.

The χ^2 test can be extended to include tables larger than the so-called 2 x 2 table of the above example. For example, a 3 x 2 table could arise if rather than classifying patients as dependent or independent, we included a third outcome category, such as "partly independent". Hence we would sum over 3 x 2 = 6 terms rather than the four terms of a 2 x 2 table. While for 2 x 2 tables the degrees of freedom is always equal to one, in general, the degrees of freedom for χ^2 tests is given by $(r-1) \times (c-1)$, where the number of rows in the table is r, and the number of columns is c.

In order for the χ^2 test to be valid, one needs to ensure that the expected values for each cell in the table is at least five. Fishers' Exact Test is often used if this criterion is not satisfied for a particular table. The Fisher's Exact test is valid for tables of any size.

Odds Ratios and Relative Risk

2

As discussed in Chapter 2, odds ratios and relative risks are often used in summarizing results of surgical research. Generically, suppose we observe the following 2 x 2 table of data,

	Disease +	Disease -	Total
Risk Factor +	a	c	$a + c$
Risk Factor -	b	d	$b + d$
Total	$a + b$	$c + d$	N

where a, b, c, and d are observed numbers of patients falling into their respective cells of the table, and $N = a + b + c + d$ is the total sample size.

Then the observed odds ratio is given by

$$OR = \frac{ad}{bc},$$

and the observed relative risk is

$$RR = \frac{\dfrac{a}{a+c}}{\dfrac{b}{b+d}}$$

Note that if the risks $\frac{a}{a+c}$ and $\frac{b}{b+d}$ are small, then $OR \approx RR$, since $a << c$ and $b << d$.

Confidence Interval for Odds Ratios

The distribution of the OR is somewhat skew, so that the confidence interval is usually based on a Normal Distribution approximation to $\log(OR)$, where log represents the natural logarithm (to the base $e = 2.71828$). In particular,

$$Var\left(\log\left(OR\right)\right) \approx \frac{1}{a} + \frac{1}{b} + \frac{1}{c} + \frac{1}{d}$$

so that a 95% CI for $\log(OR)$ is given by

$$\left(\log\left(OR\right) - 1.96 \times \sqrt{\frac{1}{a} + \frac{1}{b} + \frac{1}{c} + \frac{1}{d}},\ \log\left(OR\right) + 1.96 \times \sqrt{\frac{1}{a} + \frac{1}{b} + \frac{1}{c} + \frac{1}{d}}\right)$$

To convert back to a CI for OR, one takes the exponent (to the base $e = 2.71828$), to get

$$\left(\exp\left[\log\left(OR\right) - 1.96 \times \sqrt{\frac{1}{a} + \frac{1}{b} + \frac{1}{c} + \frac{1}{d}}\right], \exp\left[\log\left(OR\right) + 1.96 \times \sqrt{\frac{1}{a} + \frac{1}{b} + \frac{1}{c} + \frac{1}{d}}\right]\right)$$

Table 2.4. Table of χ^2 distribution probabilities. Each entry in the table provides the χ^2 value such that the probability of being greater than this value is given by the first row of the table. The number of degrees of freedom are indicated by the first column (df).

df	0.5	0.25	0.20	0.15	0.10	0.05	0.025	0.01	0.001
1	0.454	1.323	1.642	2.072	2.705	3.841	5.023	6.634	10.827
2	1.386	2.772	3.218	3.794	4.605	5.991	7.377	9.210	13.815
3	2.365	4.108	4.641	5.317	6.251	7.814	9.348	11.344	16.266
4	3.356	5.385	5.988	6.744	7.779	9.487	11.143	13.276	18.466
5	4.351	6.625	7.289	8.115	9.236	11.070	12.832	15.086	20.515
6	5.348	7.840	8.558	9.446	10.644	12.591	14.449	16.811	22.457
7	6.345	9.037	9.803	10.747	12.017	14.067	16.012	18.475	24.321
8	7.344	10.218	11.030	12.027	13.361	15.507	17.534	20.090	26.124
9	8.342	11.388	12.242	13.288	14.683	16.918	19.022	21.665	27.877
10	9.341	12.548	13.441	14.533	15.987	18.307	20.483	23.209	29.588
11	10.340	13.700	14.631	15.767	17.275	19.675	21.919	24.724	31.264
12	11.340	14.845	15.811	16.989	18.549	21.026	23.336	26.216	32.909
13	12.339	15.983	16.984	18.201	19.811	22.361	24.735	27.688	34.528
14	13.339	17.116	18.150	19.406	21.064	23.684	26.118	29.141	36.123
15	14.338	18.245	19.310	20.603	22.307	24.995	27.488	30.577	37.697
16	15.338	19.368	20.465	21.793	23.541	26.296	28.845	31.999	39.252
17	16.338	20.488	21.614	22.977	24.769	27.587	30.190	33.408	40.790
18	17.337	21.604	22.759	24.155	25.989	28.869	31.526	34.805	42.312
19	18.337	22.717	23.900	25.328	27.203	30.143	32.852	36.190	43.820
20	19.337	23.827	25.037	26.497	28.411	31.410	34.169	37.566	45.314

2

2

Similarly,

$$\text{var}(\log(RR)) \approx \frac{c}{a(a+c)} + \frac{d}{b(b+d)}$$

so that an approximate 95% CI for a RR is

$$\exp\left[\log(RR) - 1.96 \times \sqrt{\frac{c}{a(a+c)} + \frac{d}{b(b+d)}}\right], \exp\left[\log(RR) + 1.96 \times \sqrt{\frac{c}{a(a+c)} + \frac{d}{b(b+d)}}\right]$$

Going back to the example examining whether the Stroke Unit or Medical Unit is preferred, we can calculate

$$OR = \frac{67 \times 45}{34 \times 46} = 1.93$$

with 95% CI of (1.08, 3.45), and

$$RR = \frac{67/101}{46 \times 91} = 1.31$$

with 95% CI of (1.03, 1.68).

6. Statistical Inference for Means

Thus far, most of our inferential techniques have concerned dichotomous data, but similar inferential techniques are available for means. For example, referring to the data on convalescence times in Table 2.1, suppose we wish to estimate the average convalescence in each surgical group. The following formula provides 95% confidence interval limits for means (of course, the value 1.96 could be changed to other values if intervals with coverage other than 95% are of interest):

$$\left(\bar{x}_1 - 1.96\frac{s}{\sqrt{n}},\ \bar{x} + 1.96\frac{s}{\sqrt{n}}\right)$$

where $\bar{x}$ and s are the sample mean and sample standard deviation (see Section 2) from a sample of size n.

Applying this formula to the convalescence data in Table 2.1, we obtain that the mean time in the conventional surgery group is 11.1 with 95% CI of (6.2, 16.0), while the mean time in the laparoscopic group is 9.5, with 95% CI of (5.3, 13.6). While the observed mean time to convalescence is slightly shorter in the laparoscopic surgery group, the two confidence intervals are wide and largely overlap, so that we do not seem to have strong evidence from this subset of the data for an effect of surgical type on days to recovery. To be more precise, we can calculate a 95% confidence interval for the difference in means for the two groups, using the formula:

$$\left(\bar{x}_1 - \bar{x}_2 - 1.96\sqrt{\frac{s_1^2}{n_2} + \frac{s_2^2}{n_2}},\ \bar{x}_1 - \bar{x}_2 + 1.96\sqrt{\frac{s_1^2}{n_2} + \frac{s_2^2}{n_2}}\right)$$

where the notation follows that for single mean CI, with the addition of subscripts indicating the two groups. Applying that formula to our data yields a difference of

1.6 days in favor of the laparoscopic group, with 95% CI of (-4.4, 7.7) days. As expected, this interval overlaps 0, and is wide enough so that no strong conclusions could be drawn, even as to the likely direction of the effect. Further research is required (see Chapter 5 for the continuation of this example on the full data set).

As with proportions, hypothesis tests are available to supplement the confidence intervals, although we would again advise that once confidence intervals are calculated, tests add little if any additional clinically useful information. Nevertheless, for completeness, below we provide the formulae for one and two sample tests for a single mean and the difference between two means.

To test the null hypothesis that a single mean μ has value μ_0, calculate the statistic:

$$z^* = \left| \frac{\bar{x} - \mu_0}{s / \sqrt{n}} \right|$$

where $|\cdot|$ indicates the absolute value function, and determine the *p*-value from normal distribution tables as:

$$p = 2 \times Pr\{z > z^*\}.$$

For example, to test whether the average days to convalescence in the laparoscopy group is equal to $\mu_0 = 5$ days, we would calculate:

$$z^* = \left| \frac{9.46 - 5}{6.90 / \sqrt{13}} \right| = 2.33,$$

so that from Table 2.3

$$p = 2 \times Pr\{z > 2.33\} = 2 \times 0.01 = 0.02.$$

Thus we are reasonably certain that the true average value of convalescence days is not 5, in agreement with the lower limit of the CI calculated above being 6.2. Again, note that the CI provides more useful information than the *p*-value, which is not really needed once the CI is known.

The two-sample statistic to test the null hypothesis that the means in the two groups are equal to each other is:

$$z^* = \left| \frac{\bar{x}_1 - \bar{x}_2}{\sqrt{\dfrac{s_1^2}{n_1} + \dfrac{s_2^2}{n_2}}} \right|.$$

Applying this formula to test whether there is a difference between convalescence days in the two treatment groups, we find

$$z^* = \left| \frac{9.46 - 11.09}{\sqrt{\dfrac{53.69}{11} + \dfrac{47.60}{13}}} \right| = |-0.56| = 0.56.$$

Looking up 0.56 on Table 2.3 and doubling the value gives a p-value for our test of 2 x (1-.7123) = 0.5754. Hence there is no evidence in the data for a difference in mean time to convalescence, so that we cannot reject the null hypothesis that the two means are equal.

To end this section, we make the following important comments.

2

Paired versus Unpaired Tests

In comparing the two mean times to convalescence above, we have assumed that the design of this study was unpaired, meaning that the data were composed of two independent samples, one from each group. In some experiments, for example, if one wishes to compare quality of life before and after surgery is performed, a paired design is appropriate. Here one would subtract the value measured on an appropriate quality of life scale before the surgery to that measured on the same scale after the surgery to create a single set of before to after differences. Once this subtraction has been done for each patient, one in fact has reduced the two sets of before and after values on each patient to a single set of numbers representing the differences. Therefore, paired data can be analyzed using the same formulae as used for single sample analyses. Paired designs are often more efficient than unpaired designs.

Equal or Unequal Variances

The tests and confidence intervals given above assume that the variances in the two groups are unequal. Slightly more efficient formulae can be derived if the variances are the same, as a single pooled estimate of the variance can be derived from combining the information in both samples together. We do not discuss pooled variances further here, in part because in practice the difference in analyses done with pooled or unpooled variances is usually quite small, and in part because it is rarely appropriate to pool the variances, since the variability is usually not exactly the same in both groups.

Assumptions Behind the Z Tests

For ease of exposition, we have presented all of the above confidence interval and test formulae using percentiles that came from the normal distribution, but in practice there are two assumptions behind this use of the normal distribution. These assumptions are:

1. The data arise either from a normal distribution, or the sample size is large enough for the Central Limit Theorem (see Section 3.7) to apply.
2. The variance(s) involved in the calculations are known exactly.

The first of these assumptions is often satisfied at least approximately in practice, but the second assumption almost never holds in real applications. Strictly speaking, then, we should have used σ^2, σ_1^2 and σ_2^2 in the above formulae rather than s^2, s_1^2 and s_2^2, respectively, since the variances were estimated from the data rather than being known exactly. To account for the fact that the variance is estimated rather than known, we must widen our confidence intervals and increase our p-values by the appropriate amounts. It can be shown that this can be done by using t distribution tables (see Table 2.5) rather than normal distribution tables. In calculations, this means that the z^* values used in all of the above formulae need to be switched to the corresponding values from t-tables. Like the χ^2 tables, t tables require knowledge of

Table 2.5. Table of t distribution probabilities. Each entry in the table provides the t value such that the probability of being greater than this value is given by the first row of the table. The number of degrees of freedom are indicated by the first column (df).

df	0.5	0.25	0.20	0.15	0.10	0.05	0.025	0.01	0.001
1	0	1.000	1.376	1.963	3.078	6.314	12.706	31.821	318.309
2	0	0.816	1.061	1.386	1.886	2.920	4.303	6.965	22.327
3	0	0.765	0.978	1.250	1.638	2.353	3.182	4.541	10.215
4	0	0.741	0.941	1.190	1.533	2.132	2.776	3.747	7.173
5	0	0.727	0.920	1.156	1.476	2.015	2.571	3.365	5.893
6	0	0.718	0.906	1.134	1.440	1.943	2.447	3.143	5.208
7	0	0.711	0.896	1.119	1.415	1.895	2.365	2.998	4.785
8	0	0.706	0.889	1.108	1.397	1.860	2.306	2.896	4.501
9	0	0.703	0.883	1.100	1.383	1.833	2.262	2.821	4.297
10	0	0.700	0.879	1.093	1.372	1.812	2.228	2.764	4.144
11	0	0.697	0.876	1.088	1.363	1.796	2.201	2.718	4.025
12	0	0.695	0.873	1.083	1.356	1.782	2.179	2.681	3.930
13	0	0.694	0.870	1.079	1.350	1.771	2.160	2.650	3.852
14	0	0.692	0.868	1.076	1.345	1.761	2.145	2.624	3.787
15	0	0.691	0.866	1.074	1.341	1.753	2.131	2.602	3.733
16	0	0.690	0.865	1.071	1.337	1.746	2.120	2.583	3.686
17	0	0.689	0.863	1.069	1.333	1.740	2.110	2.567	3.646
18	0	0.688	0.862	1.067	1.330	1.734	2.101	2.552	3.610
19	0	0.688	0.861	1.066	1.328	1.729	2.093	2.539	3.579
20	0	0.687	0.860	1.064	1.325	1.725	2.086	2.528	3.552
21	0	0.686	0.859	1.063	1.323	1.721	2.080	2.518	3.527
22	0	0.686	0.858	1.061	1.321	1.717	2.074	2.508	3.505
23	0	0.685	0.858	1.060	1.319	1.714	2.069	2.500	3.485
24	0	0.685	0.857	1.059	1.318	1.711	2.064	2.492	3.467
25	0	0.684	0.856	1.058	1.316	1.708	2.060	2.485	3.450
26	0	0.684	0.856	1.058	1.315	1.706	2.056	2.479	3.435
27	0	0.684	0.855	1.057	1.314	1.703	2.052	2.473	3.421
28	0	0.683	0.855	1.056	1.313	1.701	2.048	2.467	3.408
29	0	0.683	0.854	1.055	1.311	1.699	2.045	2.462	3.396
30	0	0.683	0.854	1.055	1.310	1.697	2.042	2.457	3.385
40	0	0.681	0.851	1.050	1.303	1.684	2.021	2.423	3.307
50	0	0.679	0.849	1.047	1.299	1.676	2.009	2.403	3.261
60	0	0.679	0.848	1.045	1.296	1.671	2.000	2.390	3.232
70	0	0.678	0.847	1.044	1.294	1.667	1.994	2.381	3.211
80	0	0.678	0.846	1.043	1.292	1.664	1.990	2.374	3.195
90	0	0.677	0.846	1.042	1.291	1.662	1.987	2.368	3.183
100	0	0.677	0.845	1.042	1.290	1.660	1.984	2.364	3.174
200	0	0.676	0.843	1.039	1.286	1.653	1.972	2.345	3.131
500	0	0.675	0.842	1.038	1.283	1.648	1.965	2.334	3.107
1000	0	0.675	0.842	1.037	1.282	1.646	1.962	2.330	3.098
∞	0	0.674	0.842	1.036	1.282	1.645	1.960	2.326	3.299

the degrees of freedom, which for single mean problems is simply the sample size minus one. This of course applies to paired designs as well, since they reduce to single sample problems. For two sample unpaired problems, a conservative number for the degrees of freedom is the minimum of the two sample sizes minus one. These tests are thus called *t*-tests.

For example, suppose that we wanted a 95% confidence interval for a single sample problem, with a sample size of 20 patients. Rather than using the value of $z^* = 1.96$, looking at Table 2.5 with 19 degrees of freedom gives a value of 2.093 (under the 0.025 column, since leaving 2.5% on each end leaves a 95% interval). We note that the value from the *t* table is slightly larger than the corresponding value from the normal table, as expected. We also note that as the sample size increases, the difference between the values provided by the *t* table and normal table become closer to each other, so that when the degrees of freedom reaches 1000, the *t* value corresponding to the 2.5% point is 1.962, very close to the normal value. The corresponding normal values are given on the last line of Table 2.5, labeled $df = \infty$.

One-Sided or Two-Sided Test

A final issue in hypothesis testing is whether to carry out a one-sided or two-sided test. All of our tests in this section have been two-sided, which means that they have tested the null hypothesis of, for example, that a mean exactly equals zero versus the alternative hypothesis that the mean differs from zero, without specifying the direction of this difference. Most tests in the medical literature are two-sided, but on occasion one encounters one-sided tests, where the alternative hypothesis specifies either that the mean is greater than or less than the value specified in the null hypothesis.

For example, referring again to the data on days to convalescence in Table 2.1, we may specify a null hypothesis that the difference in the average number of days to convalescence is zero, versus the alternative that the days are smaller for the laparoscopic group. All of the above tests and formulae still apply, but the *p*-values are cut in half, since one considers departures from the null value in one direction only.

7. Nonparametric Inference

Thus far, statistical inferences on populations have been made by assuming a mathematical model for the population (for example, a Normal distribution), and estimating parameters from that distribution based on a sample. Once the parameters have been estimated (for example, the mean and/or variance for a Normal distribution), the distribution is fully specified. This is known as parametric inference.

Sometimes we may be unwilling to specify in advance the general shape of the distribution, and prefer to base the inference only on the data, without a parametric model. In this case, we have distribution free, or nonparametric methods.

For example, consider once again the data provided in Table 2.1.

Dividing the data for convalescent days by treatment group, we have:

Conventional surgery: 21 12 11 28 3 10 9 5 7 10 6
Laparoscopic surgery: 4 3 4 5 20 22 5 12 15 5 1 14 13

Since we are making nonparametric inferences, we no longer refer to tests of similarity of group means. Rather, the null and alternative hypotheses here are:

H_0: There is no treatment effect, i.e., laparoscopic surgery tends to give rise to convalescence days similar to those from the conventional surgery group.

H_A: Laparoscopic surgery tends to give rise to different values for convalescence days compared to those from the conventional surgery group.

In order to test these hypotheses, the first step is to order and rank the data from lowest to highest values, keeping track of which data points belong to each treatment group:

group	L	L	C	L	L	L	L	L	C	C	C	C	C
data	1	3	3	4	4	5	5	5	5	6	7	9	10
ranks	1	2	3	4	5	6	7	8	9	10	11	12	13
ranks with ties	1	2.5	2.5	4.5	4.5	7.5	7.5	7.5	7.5	10	11	12	13.5

group	C	C	C	L	L	L	L	L	C	L	C
data	10	11	12	12	13	14	15	20	21	22	28
ranks	14	15	6	17	18	19	20	21	22	23	24
ranks with ties	13.5	15	16.5	16.5	18	19	20	21	22	23	24

Thus in ranking the data, we simply sort the data from the smallest to the largest value regardless of group membership, and assign a rank to each data point depending on where its value lies in relation to other values in the data set. Hence the lowest value receives a rank of 1, the second lowest a rank of 2, and so on. Since there are many "ties" in this data set, we need to rank the data accounting for the ties, which we do by grouping all tied values together, and distributing the sum of the available ranks evenly among the tied values. For example, the second and third lowest values in this data set are both equal to 3, and there is a total of $2 + 3 = 5$ ranks to be divided among them. Hence each of these values receives a rank of $5/2 = 2.5$. Similarly, the 6th through 9th values are all tied at 5. There are $6 + 7 + 8 + 9 = 30$ total ranks to divide up amongst 4 tied values, so each receives a value of $30/4 = 7.5$, and so on.

The next step is to sum the ranks for the values belonging to the conventional surgery group, which gives:

Sum of Ranks $= 2.5 + 7.5 + 10 + 11 + 12 + 13.5 + 13.5 + 15 + 16.5 + 22 + 24 = 147.5$

We now reason as follows: There is a total of $1 + 2 + 3 + \ldots + 23 + 24 = 300$ ranks that can be distributed among the conventional and laparoscopic groups. If the sample sizes were equal, therefore, and if the null hypothesis were exactly true, we would expect that these ranks should divide equally among the two groups, so that each would have a sum of ranks of 150. Now the sample sizes are not quite equal, so that here we expect

$$300 \times \frac{11}{24} = 137.5$$

of the ranks to go to the conventional group, and

$$300 \times \frac{13}{24} = 162.5$$

of the ranks to go to the laparoscopic group. Note that $137.5 + 162.5 = 300$ which is the total sum of ranks available. We have in fact observed a sum of ranks of 147.5 in the conventional group, which is higher than expected. Is it high enough that we

2

can reject the null hypothesis? For this we must refer to computer programs that will calculate the probability of obtaining a sum of ranks of 147.5 or greater given that the null hypothesis of no treatment difference is true (remember the definition of the *p*-value, see Section 4.2). Most statistical computer packages will carry out this calculation, which in this case gives 0.58. Hence the null hypothesis cannot be rejected, as our result and those more extreme are not rare under the null hypothesis.

This nonparametric test is called the Wilcoxon rank sum test. The equivalent unpaired *t*-test for the same data also give a *p*-value of $p = 0.58$, so that the same conclusion is reached. Since the two tests do not always provide the same conclusions, which of these tests is to be preferred? The answer is situation specific. Remember that the *t*-test assumes either that the data are from a normal distribution (so here, it would imply that the days to convalescence are approximately normally distributed), or that the sample size is large. A glance back at Figure 2.1 shows that the data are skewed towards the right, so that normality is unlikely, and the sample sizes are 11 and 13, hardly large. Hence in this example the nonparametric test is preferred, since the assumptions behind the *t*-test do not seem to hold. In general, if the assumptions required by a parametric test may not hold, a nonparametric test is to be preferred, while if the distributional assumptions do likely hold, a parametric test provides slightly increased power (see Section 4.2 for the definition of statistical power) compared to a nonparametric test.

The above Wilcoxon rank sum test is appropriate for unpaired designs. A similar test exists for paired designs, called the Wilcoxon signed rank test. Nonparametric confidence intervals are also available. See Sprent[13] for details of these methods.

Pearson's and Spearman's Correlation Coefficients

Chapter 5 will cover regression methods that examine the relationship between two or more variables in detail, but here we will introduce a simple method that allows a numerical measure of the strength of the relationship between two variables, such as those displayed in Figure 2.4.

Looking at this scatter plot, there does not seem to be a strong relationship between days to convalescence and smoke years, but we would like to be more precise as to how weak or strong the relationship (if any), is. We can use the Pearson's Correlation Coefficient to measure this strength, which is calculated as follows

$$r = \frac{\sum_{i=1}^{n}(x_i - \bar{x})(y_i - \bar{y})}{\sqrt{\sum_{i=1}^{n}(x_i - \bar{x})^2 \sum_{i=1}^{n}(y_i - \bar{y})^2}} . \qquad (5)$$

The correlation coefficient ranges from -1 (perfect negative correlation, i.e., when one variable increases, the other decreases) to 1 (perfect positive correlation, i.e., when one variable increases, the other also increases), with 0 indicating no relationship.

Applied to the data from Figure 2.4 (see also Table 2.1 for the exact values that were plugged into (5), we find that the correlation $r = 0.047$, which is, as expected, a very small correlation. Confidence intervals and hypothesis tests for correlations are also available, see Armitage and Berry.[14]

Pearson's correlation measures the strength of the linear (straight line) relationship between two variables, but may not work well if the relationship is highly non linear. For example, if the relationship between two variables follows a parabolic curve, the correlation may appear to be near 0, even though a strong (but nonlinear) relationship exists. One must also be aware that even very high correlations do not necessarily imply that a causal relationship exists between two variables. For example, ice cream production may increase in the summer months, and so does the incidence of polio. However, one does not necessarily conclude that ice cream causes polio!

Spearman's correlation is a nonparametric version of Pearson's correlation, wherein one first ranks the data (separately for each variable), leading to pairs of data of ranks, rather than "raw" values. Pearson's formula (5) is then applied to the ranked data to obtain the Spearman correlation. Again, see Sprent[13] or any other book on nonparametric statistics for details.

8. Sample Size Calculations

As previously discussed, there has been a strong trend away from hypothesis testing and *p*-values towards the use of confidence intervals in the reporting of results from biomedical research. Since the design phase of a study should be in sync with the analysis that will be eventually performed, sample size calculations should be carried out on the basis of ensuring adequate numbers for accurate estimation of important quantities that will be estimated in our study, rather than by power calculations. The question of how accurate is "accurate enough" can be addressed by carefully considering the results you would expect to get (a bit of a "Catch 22" situation, since if you knew the results you will get, there would be no need to carry out the experiment!), and making sure your interval will be small enough to land in intervals numbered 1, 3, or 4 of Figure 2.8. The determination of an appropriate width is a nontrivial exercise, not to be taken lightly.

For estimating the sample requirements in experiments involving population means, two different formulae are available, depending on whether you are in a single sample or two sample situation. These are derived by solving for the sample size *n* in the formulae for the confidence intervals (see Section 6).

Single Sample

Let μ be the mean that is to be estimated, and assume that we wish to estimate μ to an accuracy of a total CI width of w (so that the CI will be $\bar{x} \pm d$, where $2 \times d = w$). Let σ be the standard deviation in the population.

Then the required sample size, n, is given by

$$n = \frac{z^2\sigma^2}{d^2} = \frac{4z^2\sigma^2}{w^2}$$

where, as usual, z is replaced by the appropriate normal distribution quantile ($z = 1.96$, 1.64, or 2.58 for 95%, 90% or 99% intervals, respectively).

For example, suppose that we would like to estimate mean systolic blood pressure postheart surgery to an accuracy of $d = 2$ mmHg with a 95% confidence interval, and that we expect the patient-to-patient variability will be $\sigma = 10$ mmHg. Then from the above formula, we need

$$n = \frac{1.96^2 \times 10^2}{2^2} = 96$$

rounding up to the next highest integer. The most difficult problem in using equation (6) is to decide on a value for the standard deviation σ, as it is usually unknown. A conservative approach would be to use the maximum value of σ that seems reasonably likely to occur in the experiment.

Two Samples

Let μ_1 and μ_2 be the means of two populations, and suppose that we would like an accurate estimate of $\mu_1 - \mu_2$. Again assume a total CI width of w (so that again $2 \times d = w$). Let σ_1 and σ_2 be the standard deviations in each population, respectively.

Then

$$n = \frac{z^2(\sigma_1^2 + \sigma_2^2)}{d^2} = \frac{4z^2(\sigma_1^2 + \sigma_2^2)}{w^2},$$

where now n represents the required sample size for each group. As usual, z is chosen as above.

Similar formulae are available for sample size requirements for accurate estimation of one or two proportions.

Single Sample

Let p be the proportion that is to be estimated, and assume that we wish to estimate p to an accuracy of a total CI width of $w = 2 \times d$.

Then

$$n = \frac{z^2}{d^2} p(1-p) = \frac{4z^2}{w^2} p(1-p),$$

where, again, z is the appropriate normal quantile.

Two Samples

Let p_1 and p_2 be the two proportions whose difference we would like to estimate to a total CI width of $w = 2 \times d$.

Then

$$n = \frac{4(p_1(1-p_1) + p_2(1-p_2))z^2}{w^2} = \frac{(p_1(1-p_1) + p_2(1-p_2))z^2}{d^2} \qquad (8)$$

where again n represents the required sample size for each group.

For example, suppose we would like to design a study to measure the difference in success rates for two types of surgery. Assume that the standard therapy is thought to be successful with probability $p_1 = 0.70$, and the new surgery may improve this to $p_2 = 0.80$. We would like to estimate the true difference to within $d = 0.05$, so that not only will we be able to detect any differences of 10%, but the 95% confidence interval will be far enough away from 0 (if our predicted rates are correct) so that we can make a more definitive conclusion as to the clinical utility of the new technique (see Section 4.3). We calculate

$$n = \frac{(p_1 \times (1-p_1) + p_2 \times (1-p_2)) \times z^2}{d^2} = \frac{(0.7 \times (1-0.7) + 0.8 \times (1-0.8)) \times 1.96^2}{0.05^2} = 569$$

so that 569 patients are required in each group.

The main practical difficulty with formulae (7) and (8) is assigning appropriate values for p, p_1 and p_2. It is therefore useful to note that equation (7) is maximized when $p = 0.5$, so using this value is conservative in the sense that the desired confidence interval width will be respected regardless of the estimated value of p that will be observed in the study. This conservative value, however, may provide too large a sample size and therefore be wasteful of resources if the true proportion is far from 0.5. A conservative rule of thumb is to use the value of p which is closest to 0.5, selected from the set of all plausible values. Similarly, (8) is maximized for $p_1 = p_2 = 0.5$, so a similar rule of thumb applies for each of p_1 and p_2.

9. Bayesian Inference

Consider again the problem introduced in Section 4.1. Recall that in this example the standard therapy is assumed to have a success rate of 80%, while the data collected so far for the new surgical technique indicates a 90% success rate, but is based on only 10 subjects. The frequentist confidence interval was very wide, ranging from 55.7-99.7%, so has not been particularly helpful in making a decision as to which technique to use for the next patient. At this point, with the data being relatively uninformative, the surgeon may decide to be conservative and remain with the standard surgery until more information becomes available about the new technique, or may go with his "gut feeling" as to the likelihood that the new therapy is truly better or not. If there has been data from animal experiments or strong theoretical reasons why the new technique may be better, he may be tempted to try the new one. Can anything be done to aid him in this decision making process?

Bayesian analysis has several advantages over standard or frequentist statistical analyses. These advantages include:

1. The ability to address questions of direct clinical interest, such as questions 1 and 2 of Section 4.1. Hence results of Bayesian analyses are straightforward to interpret, in contrast to the obscure and difficult to understand (and hence frequently misinterpreted) inferences provided by *p*-values and confidence intervals.
2. The ability to incorporate relevant information not directly contained in the data into any statistical analysis.
3. A natural way to update statistical analyses as new information becomes available.

A main theoretical difference between frequentist and Bayesian statistical analyses is that Bayesian analysis permits parameters of interest (binomial probabilities, population means, and so on) to be considered as random quantities, so that probabilities can be attached to the possible values that they may attain. On the other hand, frequentists consider these parameters as fixed (albeit possibly unknown) constants, so have no choice but to attach their probabilities to the data that could arise from the experiment, rather than on the parameters. This distinction is the main why Bayesian analysis is able to answer the direct questions of interest (questions 1 and 2 of Section 4.1) while frequentist analyses must settle for answering the more obscure questions 3 and 4.

The ability to address questions of direct interest, however, comes at the cost of having to do a bit more work. Not only do Bayesians have to collect data from their

2

experiments, but they also have to quantify the state of knowledge about all parameters prior to their collecting this data. This nontrivial step is summarized in a prior distribution. The information in the prior distribution is updated by the information in the data to arrive at a posterior distribution, which summarizes all available information, past and current. We will apply a Bayesian analysis to our surgeon's decision later in this section, but first we need to define the basic elements of all Bayesian analyses, and see how they are applied to drawing inferences about our parameter of interest here, the binomial success rate of the new surgical technique.

Let us generically denote our parameter of interest as θ. Hence θ can be a binomial parameter, or the mean and variance from a Normal distribution, or an odds ratio, or a set of regression coefficients, and so on. Note in particular that θ can be two or more dimensional. The parameter of interest is sometimes usefully thought of as the "true state of nature". The basic elements of a Bayesian analysis then are:

1. The prior probability distribution, $f(\theta)$. This prior distribution summarizes what is known about θ before the experiment is carried out. It is "subjective", so may vary from investigator to investigator.
2. The likelihood function, $f(x|\theta)$. The likelihood function provides the distribution of the data, x, given the parameter value θ. For instance, it may be a binomial likelihood (see equation (1)), a normal likelihood of the form given in equation (2), or a likelihood from a regression equation with associated normal residual variance (see Chapter 5). It is important to realize that Bayesians and frequentists alike can use the same likelihood function, as both need to calculate the probability of data given various values for the parameter θ.
3. The posterior distribution, $f(\theta \mid x)$. The posterior distribution summarizes the information in the data, x, together with the information in the prior distribution, $f(\theta)$. Thus, it summarizes what is known about the parameter of interest θ after the data are collected.

Bayes Theorem, first discussed by Thomas Bayes[15] in 1763 relates the above three quantities:

$$\text{posterior distribution} = \frac{\text{likelihood of the data} \times \text{prior distribution}}{\text{a normalizing constant}},$$

or using our notation above,

$$f(\theta|x) = \frac{f(x|\theta) \times f(\theta)}{\text{a normalizing constant}}$$

or, omitting the normalizing constant,

$$f(\theta|x) \propto f(x|\theta) \times f(\theta)$$

where $\propto$ indicates "is proportional to".

Thus we "update" the prior distribution to a posterior distribution after seeing the data via Bayes Theorem. The current posterior distribution can be used as a prior distribution for the next study, hence Bayesian inference provides a natural way to represent the learning that occurs as science progresses.

The most contentious element in Bayesian analysis is the need to specify a prior distribution. Since there is no "objective" or unique way to derive prior distributions, they are necessarily subjective, in the sense that one surgeon may derive a different prior distribution than his colleague, and hence arrive at a different posterior distribution.

Several points can be made regarding this "controversy":

1. Bayesian can use "diffuse", "flat" or "reference" prior distributions, which for all practical purposes consider all values in the feasible range as equally likely. Hence, if little prior information exists, or if a Bayesian wishes to see what information the data themselves provide, this choice of prior distribution can be used. In fact, in many situations, a Bayesian analysis using reference priors will result in similar interval estimates as those provided by frequentist confidence intervals. However, the Bayesian intervals still retain their nice interpretation as directly providing the probability that the parameter of interest will be in the interval, as compared to the somewhat convoluted definition of the frequentist confidence interval.
2. While many frequentists have been quick to criticize Bayesian analysis because of the difficulty in deriving prior distributions, frequentist analysis always completely ignores this information, which can hardly be considered as a better solution.
3. If different clinicians have a range of prior opinions and hence a range of prior distributions, there will also be a range of posterior distributions. Presenting several Bayesian analyses matching this range of prior opinions helps to raise the level of debate following the publication of results in medical journals, as it accurately reflects the range of clinical opinion that exists in the community. Furthermore, it can be shown that as more data accumulates, the posterior distributions from different priors tend to converge towards a single distribution, accurately mirroring the process of eventual consensus among clinicians as data accumulates. See Speigelhalter et al[16] or a more introductory level article in JAMA[10] for more information on using a range of prior distributions when carrying out a Bayesian analysis.

9.1. Bayesian Inference for Proportions

Suppose that in a given experiment x "successes" are observed in N binomial trials. Let θ denote the true but unknown probability of success, and suppose that the problem is to find an interval that covers the most likely locations for θ given the data. We use the notation θ in keeping with our generic notation above, but note that θ here is equivalent to p from Sections 3 and 4.

The Bayesian solution to this problem follows the usual pattern, as outlined above. Hence the main steps can be summarized as:

1. Write down the likelihood function for the data.
2. Write down the prior distribution for the data.
3. Use Bayes Theorem (that is, multiply the equation for the likelihood function of the data by the prior distribution) to derive the posterior distribution. Use this posterior distribution, or summaries of it like 95% credible

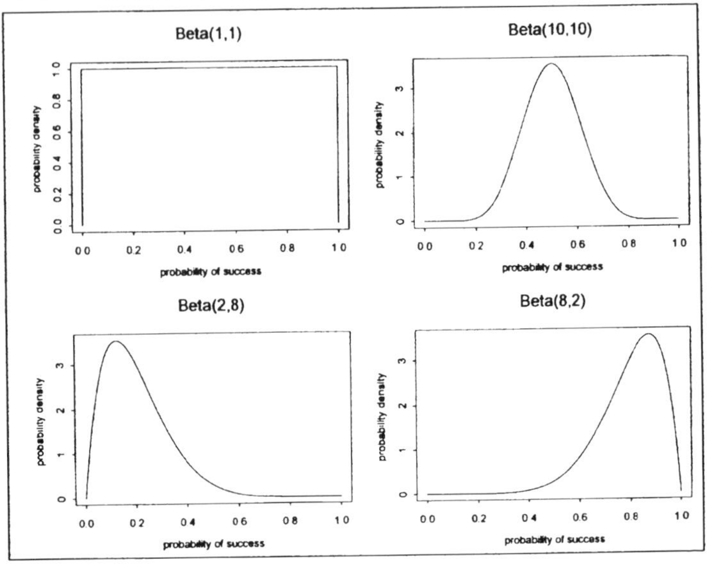

Fig. 2.9. Four beta densities. The distribution in the upper left hand corner is a beta(α=1, β = 1) density, equivalent to a uniform distribution. The distribution in the upper right hand corner is a beta(α =10, β = 10) density, which looks similar to a normal distribution in shape. The bottom left and right densities are beta(α = 2, β = 8) and beta(α =8, β = 2) densities, respectively, which are both skewed.

intervals for statistical inferences. Credible intervals are the Bayesian analogues to frequentist confidence intervals.

For the case of a single binomial parameter, these steps are realized by:

$$f(\theta|x) = Pr\{x \text{ success in } N \text{ trials}\} = \frac{N!}{(N-x)!x!}\theta^x(1-\theta)^{(N-x)},$$

1. The likelihood function is the usual binomial probability formula given by equation (1) in Section 3.4,
   where $f(\theta|x)$ represents the likelihood function for the success rate θ given data x.
2. Although any prior distribution can be used, for reasons to be explained below, a convenient prior distribution family is the beta family. A random variable, θ, has a distribution that belongs to the beta family if it has a probability density given by:

$$f(\theta) = \frac{1}{B(\alpha,\beta)}\theta^{\alpha-1}(1-\theta)^{\beta-1}$$

for $0 \leq \theta \leq 1$, and α, $\beta > 0$. [B(α,β) represents the beta function evaluated at (α,β). It is simply the normalizing constant that is necessary to make the total area under the curve to one, but otherwise plays no role.]
Some beta distributions are illustrated in Figure 2.9. For example, using

a beta($\alpha = 1$, $\beta = 1$) distribution produces a perfectly flat or uniform distribution over the range of all possible values, suitable for a "diffuse" or "noninformative" distribution when little or no prior information is available or when one wishes to see the information contained in the data by themselves. On the other hand, a beta($\alpha = 10$, $\beta = 10$) density produces a curve similar in shape to a normal density centered at $\theta = 0.5$. If $\alpha > \beta$ then the curve is skewed towards values near 1, while if $\alpha < \beta$ then the curve is skewed towards values near 0.

The mean of the beta distribution is given by:

$$\mu = \frac{\alpha}{\alpha + \beta},$$

and the standard deviation is given by:

$$\sigma = \sqrt{\frac{\alpha\beta}{(\alpha+\beta)^2(\alpha+\beta+1)}}.$$

Therefore, to choose a prior distribution, one needs only to specify values for α and β. This can be done by finding the α and β values that give the correct prior mean and standard deviation values. Solving the above two equations in two unknowns, the formulae are:

$$\alpha = \frac{-\mu\left(\sigma^2 + \mu^2 - \mu\right)}{\sigma^2}$$

and

$$\beta = \frac{(\mu - 1)\left(\sigma^2 + \mu^2 - \mu\right)}{\sigma^2}.$$

For example, if we wish to find a member of the beta family centered at $\mu = 0.85$ and with $\sigma = 0.05$, then plugging these values for μ and σ into the above equations gives $\alpha = 42.5$ and $\beta = 7.5$, so that a $\beta(42.5, 7.5)$ will have the desired properties. This curve, pictured in Figure 2.10, may be an appropriate prior distribution for the problem introduced in Section 4.1, if the surgeon believes, a priori, that the new technique is likely to be successful between 75-95% of the time, and whose best guess of the rate is 85%. We will return to this example again shortly.

3. As always, Bayes Theorem says:

   posterior distribution $\propto$ prior distribution $\times$ likelihood function.

   In this case, it can be shown (by relatively simple algebra) that if the prior distribution is beta(α,β), and the data is x successes in N trials, then the posterior distribution is beta($\alpha + x$, $\beta + N - x$). This simplicity arises from noticing that both the beta prior distribution as represented in (9) and the binomial likelihood as given in (1) have the general form $\theta^{\alpha} \times (1-\theta)^{\beta}$, so that when multiplying them as required by Bayes Theorem, the exponents simply add, and the form is once again recognized to be from the beta family of distributions.

2

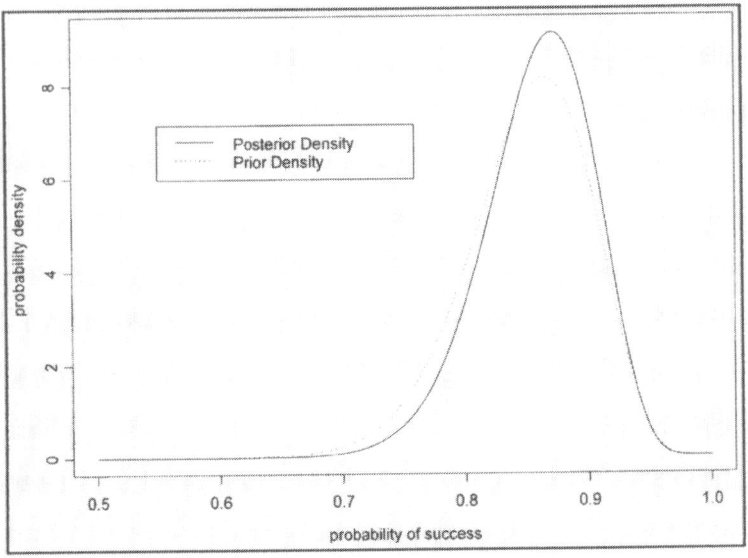

Fig. 2.10. Prior and posterior densities for the binomial probability example.

Hence if we observe the new surgical technique to be successful in 9 out of 10 patients, and if we use the prior distribution discussed above, then the posterior distribution is a beta(42.5 + 9, 7.5 + 1) = beta(51.5, 8.5) distribution, which is illustrated in Figure 2.10. The mean of this distribution is $\frac{51.5}{51.5+8.5} = 0.858$, and the 95% posterior interval is (0.77, 0.94). The probability of being greater than 80% is 0.896 (area under the curve to the right of 0.8 in Figure 2.10). Therefore, the surgeon may be tempted to try the new surgery on his next patient, but should realize that this decision is mostly based on his prior information, to which the data contributed only a small amount of new information. Looking at Figure 2.10, we see that the prior density was shifted only a small amount by the data. If instead he "lets the data speak for themselves" by using a beta(1,1) or uniform prior distribution (see Figure 2.9), then the 95% interval is (0.63, 0.99), very similar numerically to the frequentist confidence interval, although their interpretations are very different. Bayesian intervals (often called credible intervals to distinguish them from frequentist confidence intervals) are interpreted directly as the posterior probability that θ is in the interval, given the data and the prior distribution. No references to long run frequencies or other experiments are required, as is the case for confidence intervals (see Section 4.3). Of course, the chart in Figure 2.8 applies to Bayesian credible intervals as well as frequentist confidence intervals.

In general, one should usually carry out a Bayesian analysis using a "diffuse" prior distribution like a beta(1,1) distribution, to examine what information the current data set provides. Then, one or more Bayesian analyses with more informative prior distributions could be carried out, depending on the available prior information. If there are widely divergent opinions in the medical community concerning the parameters of interest, then several prior distributions should be used. If the data set is large, then similar conclusions will be reached no matter which prior distribution one starts with. On the other hand, with smaller data sets, there will still be diversity of opinions, even after the new data are analyzed. Bayesian analysis allows this situation to be accurately represented and assessed.

Bayesian Inference for Means

Example

Consider the situation where we are trying to estimate the average age of men who arrive at a certain emergency clinic with a myocardial infarction (MI).

Suppose that the following data are collected on 27 consecutive such men at this clinic:

76, 71, 82, 63, 76, 64, 64, 74, 70, 64, 75, 81, 75, 78, 66, 62, 79, 82, 78, 62, 72, 83, 79, 41, 80, 77, 67.

From this data, we find $\bar{x} = 71.89$, and $s^2 = 85.18$, so that $s = \sqrt{85.18} = 9.22$.

Let us assume the following:

1. The standard deviation is known a priori to be $\sigma = 9$ years. This unrealistic assumption allows us to forgo estimating the variance, which complicates the analysis somewhat. Of course, Bayesian methods for analyzing data when the standard deviation is unknown are available, see the book by Berry.[17]
2. The observations come from a Normal distribution with unknown mean μ (and known variance $\sigma^2 = 9^2 = 81$).

We will again follow the three usual steps used in Bayesian analyses:

1. Write down the likelihood function for the data.
2. Write down the prior distribution for the unknown parameter, in this case the normal mean μ.
3. Use Bayes Theorem to derive the posterior distribution. Use this posterior distribution, or summaries of it like 95% credible intervals for statistical inferences.

For inference about a normal mean, μ, these steps are:

1. The likelihood function for the data is based on the Normal distribution, i.e.,

$$f(\mu|x) = \prod_{i=1}^{n} \frac{1}{\sqrt{2\pi}\sigma} \exp\left(-\frac{(x_i - \mu)^2}{2\sigma^2}\right) = \left(\frac{1}{\sqrt{2\pi}\sigma}\right)^n \exp\left(\frac{\sum_{i=1}^{n}(x_i - \mu)^2}{2\sigma^2}\right).$$

   This likelihood function is derived from multiplying together n copies of the normal density, each time plugging in the data, x_i, from a single patient (see equation (2)). Hence the likelihood function represents the probability of getting the observed data, where the data follow a normal curve. In our example, $n = 27$.

2. Suppose that we have a priori information that the random parameter μ is likely to be in the interval (60,80). That is, we think that the average

age should be about 70, but would not be too surprised if it were as low as perhaps 60, or as high as about 80. We will represent this prior distribution as a second normal distribution (not to be confused with the fact that the data are also assumed to follow a Normal density). The normal prior density is chosen here for the same reason as the Beta distribution is chosen when we looked at the binomial distribution: it makes the solution of Bayes Theorem very easy. We can therefore approximate our prior knowledge as:

$$\mu \sim N(\theta, \tau^2) = N(100, 5^2 = 25),$$

where $\tau = 5$ is our prior standard deviation, and where the notation $\sim$ is read as "is distributed as". Small values of τ would indicate that we are quite certain that the true mean is near 70 years, while larger values would indicate less certainty about our prior mean value of $\theta = 70$.

In general, this choice for a prior distribution is based on any information that is available at the time of the experiment. In this case, the prior distribution was chosen to have a somewhat large standard deviation ($\tau = 5$) to reflect that we are quite uncertain about the average age of the MI patients. A clinician with more experience in this area, or who perhaps has worked in similar clinics for a long period of time, may elect to choose a much smaller value for τ. Note that we use two distinct standard deviations: σ represents the variability of the ages among the patients, while τ represents how certain we are of our prior mean value.

3. We now wish to combine this prior density with the information in the data as represented by the likelihood function to derive the posterior distribution. This combination is again carried out by a version of Bayes Theorem. Hence we multiply the likelihood function by the prior density. After some algebra, the posterior distribution is again given by a normal distribution,

$$N\left(A \times \theta + B \times \bar{x},\ \frac{\tau^2 \sigma^2}{n\tau^2 + \sigma^2}\right),$$

where

$$A = \frac{\sigma^2 / n}{\tau^2 + \sigma^2 / n} = 0.107,\ B = \frac{\tau^2}{\tau^2 + \sigma^2 / n} = 0.893,\ n = 27,\ \sigma = 9,\ \tau = \sqrt{25} = 5,\ \theta = 70, \text{ and } \bar{x} = 71.89.$$

Plugging in these values we find that the posterior distribution for our mean μ is $N(71.69, 2.68)$, as shown in Figure 2.11. The mean value depends on both the prior mean, θ, and the observed mean, $\bar{x}$.

Once again, the posterior distribution is interpreted as the actual probability density of μ given the prior information and the data, so that we can calculate the probabilities of being in any interval we like. These calculations can be done in the usual way, using normal tables. For example, a 95% credible interval is given by (68.5, 74.9). In comparing this interval to the prior 95% interval of (60,80), we can see that the data here have provided much information. In fact, if we use a "flat" or

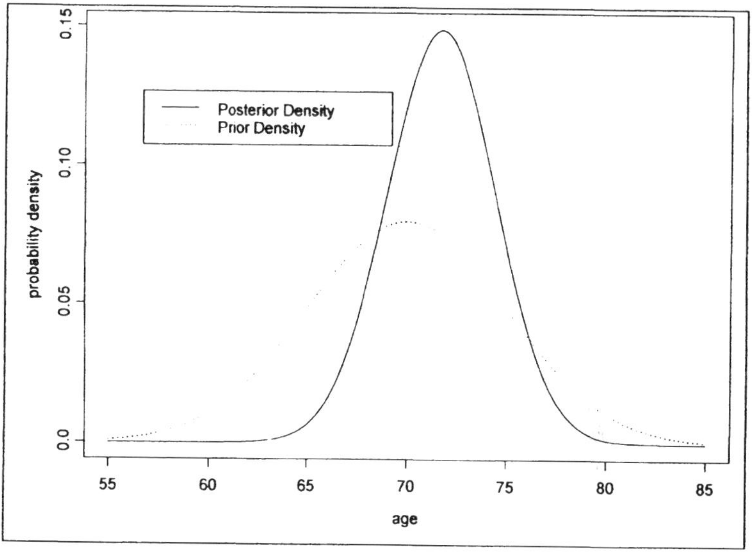

Fig. 2.11. Prior and posterior densities for the age of MI example.

diffuse prior distribution to see what the data themselves say, the 95% credible interval is (68.3, 75.1), which is hardly different from the previous interval.

Summary of the Use of Bayesian Inference versus Frequentist Inference

In this Chapter, we have considered both standard frequentist analysis and Bayesian analysis. In general, Bayesian analysis is more difficult to carry out, since in addition to collecting some data, one needs to carefully assess a prior density. In any given situation, which is preferable? The following points should be considered in making this decision:

1. Bayesian analysis allows for the integration of past knowledge, presumably leading to sharper inferences and better clinical decisions when other information is relevant to the decision, as was the case in our example of Section 9.2. See Brophy and Joseph[10] for an example using data from a clinical trial. Many other examples are also available.[11,16,17]
2. It is more work to carry out Bayesian analyses. This extra effort may be worthwhile for the questions that Bayesian analysis allows one to address, or if there is appreciable prior information that should be factored into any decisions. While frequentists are also able to consider prior information, they can do so only informally. This usually appears as nonquantitative ad hoc discussion, which is difficult for readers to evaluate.
3. Frequentist analysis is at the present time more commonly used in the medical literature compared to Bayesian analysis, but that does not necessarily imply it is better or that is to be preferred in any given situation.

The use of Bayesian analysis has been rapidly increasing in the past few years, and it remains to be seen what the main paradigm of statistical analysis will be in the 21st century. There is usually a time lag in bringing innovations from one scientific domain into another, and this has been especially problematic in bringing new statistical methods into the medical literature.

Conclusions

This Chapter has introduced some of the major ideas behind statistical inference, with emphasis on the simpler methods most useful to medical research. Rather than a simple catalogue listing of which tests to use for which types of data, we have tried to explain the logic behind the common statistical procedures seen in the medical literature, the correct way to interpret the results, and what their drawbacks may be. We have also introduced Bayesian inference as a strong alternative to standard frequentist statistical methods, both for its ability to incorporate the available prior information into the analysis, and because of its ability to address questions of direct clinical interest.

Of course, there are many other important statistical techniques used in medical research. The three most important of these are linear and logistic regression, both covered in Chapter 5 of this book, and survival analysis, discussed in Chapter 6. Statistical issues related to diagnostic testing are addressed in Chapter 3, and Chapter 7 provides an introduction to meta-analysis, which is the statistical merging of information from several studies.

For further reading, there are literally hundreds of books on basic statistical techniques, with many of these specific to statistics in medicine. These include Colton,[18] Armitage and Berry,[14] and Rosner.[19] Berry's book[17] provides an introduction to Bayesian methods, and this same author also has coedited a book on more advanced Bayesian applications to medicine.[20]

Selected Readings

1. Savage M, Douglas J, Fischman D et al. Stent placement compared with balloon angioplasty for obstructed coronary bypass grafts. N Eng J Med 1997; 740-747.
2. Savage L. The subjective basis of statistical practice. Technical Report, Department of Statistics, University of Michigan, Ann Arbor, 1961.
3. Berger J. Statistical Decision Analysis and Bayesian Analysis. Second Edition. New York: Springer Verlag 1985.
4. Rosa L, Rosa E, Sarner L et al. A close look at therapeutic touch. J Amer Med Assoc 1998; 279:1005-1010.
5. Royall R. Statistical Evidence. New York: Chapman and Hall 1997.
6. Demptster A. Bayes, Fisher, and belief functions. In: Geisser S, Hodges J, Press S et al, eds. Bayesian and likelihood methods in statistics and econometrics: Essays in honor of George A. Barnard. North Holland: Elsevier 1990.
7. Box G. Statistics for experimenters: An introduction to design, data analysis, and model building. New York: Wiley 1978.
8. Rothman K. Writing for Epidemiology. Epidemiology 1998:9.
9. Evans S, Mills P, Dawson J. The end of the p-value. British Heart Journal 1988; 60:177-180.
10. Brophy J, Joseph L. Placing trials in context using Bayesian analysis: GUSTO revisited by reverend Bayes. J Amer Med Assoc 1995; 273(11):871-875.

11. Lilford R, Braunholz D. The statistical basis of public policy: A paradigm shift is overdue. Brit Med J 1996; 313:603-607.
12. Garraway W, Akhtar A, Prescott R et al. Management of acute stroke in the elderly: Preliminary results of a controlled trial. Brit Med J 1980; 280:1040-3.
13. Sprent P. Applied nonparametric statistical methods. Chapman and Hall, New York, 1989.
14. Armitage P and Berry G. Statistical Methods in Medical Research, Third Edition. Oxford: Blackwell Scientific Publications 1994.
15. Bayes, T. An essay towards solving a problem in the doctrine of chances. Philosophical Transactions of the Royal Society. 1763; 53:370-418.
16. Spiegelhalter D, Freedman L, Parmar M. Bayesian approaches to randomized trials. Journal of the Royal Statistical Society, Series A 1994; 157:387-416.
17. Berry D. Statistics—A Bayesian perspective. Duxbury, Belmont, 1996.
18. Colton T. Statistics in Medicine. Little Brown, Boston, 1974.
19. Rosner B. Fundamentals of Biostatistics. Duxbury, Belmont, 1995.
20. Berry D, Stangl D eds. Bayesian Biostatistics. Marcel Dekker, New York, 1996.

2

CHAPTER 3

Interpretation of Diagnostic Tests

Joseph Romagnuolo, Alan N. Barkun and Lawrence Joseph

After history and physical examination, the clinician is often left with a short differential diagnosis that must be further narrowed before appropriate treatment can begin. This almost always means that additional radiographic and/or laboratory tests must be performed in order to confirm or discard a suspected diagnosis. In fact, even the questions asked of patients during the history and physical exam are all informal diagnostic tests in themselves. How likely is the patient to have a mass if you don't feel one? How likely is the patient with recurrent biliary colic and normal transaminases to have a common bile duct stone, and should they have an ERCP before their cholecystectomy? How confident can you be that your patient with a normal barium enema does not have polyps or cancer? What if they had had a negative sigmoidoscopy as well? The interpretation of the literature studying these tests and the translation of that information into a form that is easily applicable to a clinical scenario is thus critical. In this review, we will introduce several statistical concepts relating to diagnostic tests and provide examples demonstrating how these can be used in day to day practice.

Diagnostic Tests and the 2 x 2 Table

The 2x2 ("two by two") table is a useful way of summarizing data regarding the performance of a diagnostic test. The name arises from the form of the table, which has 2 rows and 2 columns. The 2 columns represent patients with (D+) and without (D-) the disease sought, and the rows are the patients with a positive test (T+) and a negative or normal test (T-). The columns and rows are totaled at the bottom and right side, respectively (Table 3.1). The values (numbers of patients) falling into the four cells in Table 3.1 are each represented by a letter (a, b, c, d) to facilitate the referencing of these cells in the discussion below.

The study represented in Table 3.1 concerned the use of ultrasound in 70 children to diagnose acute appendicitis.[1] A positive test was defined as a noncompressible appendix with a maximal outer diameter greater than 6 mm. Thirty-five patients had a positive test, and of those, 31 had pathologically documented appendicitis while 4 settled without therapy. It is assumed that the latter did not have appendicitis (false positive patients). Of the 33 with acute appendicitis, two patients were missed (false negative patients).

Surgical Arithmetic: Epidemiological, Statistical and Outcome-Based Approach to Surgical Practice, by Lawrence Rosenberg, Lawrence Joseph, and Alan Barkun. ©2000 Landes Bioscience.

Table 3.1. Acute appendicitis (Data from Vignault et al)

		D+	D-	
Ultrasound	T+	a = 31	b = 4	a+b = 35
Finding	T-	c = 2	d = 33	c+d = 35
		a+c = 33	b+d = 37	a+b+c+c = 70

One can easily compute clinically relevant proportions from such a table. For example, 2 of 33 patients with acute appendicitis had a negative test for a false negative rate of 6%. The probability of any patient having a positive test can be written in shorthand as P(T+), which can be estimated from the table as $(a+b)/(a+b+c+d) = 35/70 = 0.50$. Of course, we know from Chapter 2 that this estimate is subject to random error, and that calculating a confidence interval for this probability is a good idea. Here the 95% confidence interval is (0.38,0.62). A patient can either have a positive or a negative test but not both so that $P(T+) + P(T-) = 1$. Similarly, one can either have appendicitis or not ($P(D+) + P(D-) = 1$).

Conditional probability is the probability that an event occurs given that another condition is satisfied or that another event occurs, for example, the probability that a patient does not have appendicitis given that he/she has had a positive test. This is notated as $P(D-|T+)$, which can be estimated from the 2x2 table as $4/35 = b/(a+b) = 0.11$. The numerator is the figure in the D-/T+ box and the denominator is the total from the T+ row. See Chapter 2 for more details about probability, conditional probability, and confidence intervals.

Sensitivity and Specificity vs Predictive Values

Sensitivity, specificity, positive and negative predictive values, and accuracy are often collectively referred to as characteristics of test performance. Sensitivity and specificity are often said to be inherent to a test, whereas predictive values vary depending on the prevalence of disease (or alternatively, the pretest probability of disease) in the population tested. This important distinction will be examined further below. We first define each of these quantities.

The sensitivity of a test is the conditional probability of a patient testing positive for a disease given that the patient in fact has the disease. Referring to Table 3.1, we can estimate the sensitivity of ultrasound as $P(T+|D+) = 31/33 = a / (a+c) = 0.94$. Note that the denominator is the total number of patients with the disease. In a perfectly sensitive test, all patients with the disease test positive. It is also known as the true positive rate or PiD (positive in disease) rate.[2] As such, a sensitive test is helpful at ruling OUT disease. This can be remembered by the mnemonic SnOUT (Sn for sensitivity). This is the characteristic one looks for in an initial screening test or when the disease is so devastating that one cannot afford to miss it. The sensitivity is also 1-(false negative rate) or $1-P(T-|D+)$ which in our example can be estimated as $1-c / (a+c) = 1-2/33 = 31/33 = 0.94$, as above.

The specificity of a test is the probability of a patient testing negative for a disease given that the patient is in fact disease-free. Again, referring to Table 3.1, the specificity of ultrasound can be estimated by $P(T-|D-) = d / (b+d) = 33/37 = 0.89$.

Note that the denominator is the number of nondiseased patients. In other words, the specificity provides the probability that normal patients will have a normal test. It can also be thought of as the true negative rate or the NiH (negative in health) rate. A perfectly specific test allows one to, therefore, rule IN disease with absolute certainty because you are sure that you have correctly classified everyone who is nondiseased. The mnemonic SpIN is often used.[3] A specific test is useful as a confirmatory test, especially when a positive test may lead to a potentially dangerous therapeutic intervention. As well, tests with poor specificity may lead to increased health care expenditures because of unnecessary further testing or therapeutic interventions. Specificity is also 1-(false positive rate) or $1 - P(T+|D-)$, which in our ultrasound example is estimated by $1\text{-}b/(b+d) = 1\text{-}4/37$. Unfortunately, when most tests are manipulated to increase sensitivity, for example by decreasing the critical appendiceal diameter on ultrasound from 6 mm to 4 mm, this will almost always at the same time decrease specificity because one is now calling some normal appendices abnormal. Similarly, attempts to increase specificity by raising a cutoff value usually results in decreased sensitivity. This inverse relationship, illustrated in Figure 3.1, has led to the development of test characteristics that incorporate both sensitivity and specificity in one, such as the likelihood ratio, discussed later.

Unlike sensitivity and specificity, which have as their "denominators" the number of people with or without a disease, respectively, predictive values consider the number of people with or without a positive test. Positive predictive value (PPV) is defined as the conditional probability of in fact having the disease given a positive test result. Referring again to Table 3.1, we can estimate the PPV of ultrasound for the diagnosis of acute appendicitis as $P(D+|T+) = a/(a+b) = 31/35 = 0.89$. Predictive values are felt to be the most clinically relevant test characteristic since they provide information about the type of situation clinicians face most often: the physician has received a report of a positive test result and wants to know how likely the disease now is. PPV is another way of measuring the ability of a test to "rule in" disease, and so the mnemonic SpIN can be modified to SpPin, where the capital "P" in the center stands for PPV.[2]

However, because sensitivity and specificity consider only either diseased or normal subjects, respectively, and do not mix these groups, they are not influenced mathematically by changes in the prevalence of disease in the tested population. The prevalence of disease in the above example can be estimated as $P(D+) = (a+c)/(a+b+c+d) = 33/70 = 0.47$. If we were to test a population with a lower pretest probability of disease, for example, and halve all the values in the D+ column of Table 3.1 (see Table 3.2), then the sensitivity and specificity would not change, because the proportions of patients who test positively or negatively in the D+ and D- columns, respectively, would not change. However, the PPV will drop from 0.89 to $15.5/35 = 0.44$. Therefore, the PPV goes down when the prevalence of disease goes down and goes up when specificity goes up. The corollary is that PPV can be high in a population with high disease prevalence even if the specificity is poor.

Similarly, negative predictive value (NPV) is the probability of a patient being truly nondiseased given that the test is normal. NPV can be estimated by $P(D-|T-) = d/(c+d) = 33/35 = 0.94$ in Table 3.1, and as $33/34 = 0.97$ in Table 3.2. The NPV improves as disease prevalence goes down, in direct contrast to the PPV. It is also higher the more sensitive a test is. The NPV is another measure of the ability of a

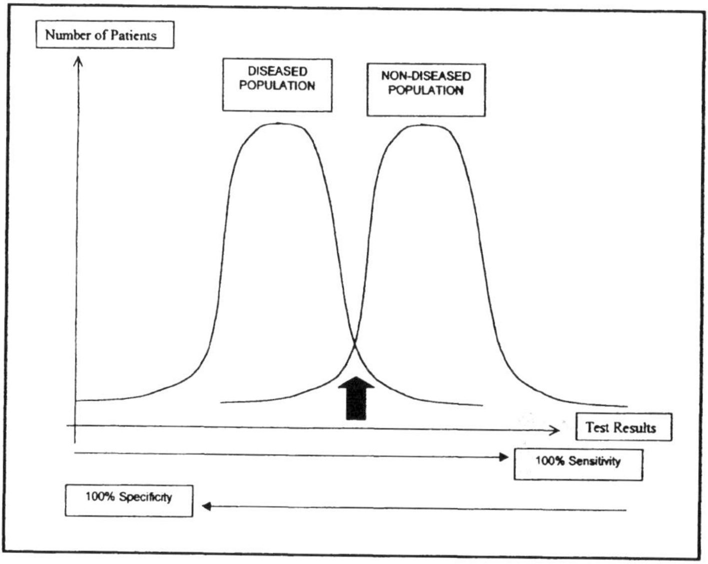

Fig. 3.1. Overlapping distributions of diseased and nondiseased populations. As the cutoff value is moved to the right, sensitivity increases and specificity decreases. As it is moved to the left, sensitivity decreases while specificity increases. The black block arrow indicates the point of minimum overlap whereon a cutoff value would yield the best mathematical compromise between sensitivity and specificity. The best clinical compromise may be at a different point.

Table 3.2. Acute appendicitis

		D+	D-	
ultrasound	T+	a = 15.5	b = 4	a+b = 19.5
finding	T-	c = 1	d = 33	c+d = 34
		a+c = 16.5	b+d = 37	a+b+c+c = 53.5

test to "rule out" disease. The SnOUT mnemonic can also be modified to SnNout where the capital "N" in the center stands for NPV.[2]

Although the classical epidemiological model for diagnostic tests discussed above implies that sensitivity and specificity are inherent properties of a test and do not change when prevalence of disease does, this is in reality an oversimplification (albeit a very useful one). When tests are applied to differing populations or when the diagnostic test is applied to a heterogeneous group of disorders, or a heterogeneous group of patients within a given disorder, there is always the possibility of spectrum bias.[4-6] The sensitivity of a test in identifying advanced disease, for example, is frequently different when the same test is used for identifying early disease. The sensitivity of barium enema for the detection of polyps, in theory, should not change whether you are looking at 50 year olds with a low prevalence of neoplasia or 70 year olds with a higher prevalence. Unfortunately, the sensitivity is affected by factors such as the size of the polyps. And since the higher prevalence group will also have the larger, more obvious lesions, the sensitivity of the test will appear to go up in the higher prevalence group. Therefore, if the disease in the higher prevalence group represents a different spectrum of the disease, that is, with cases that are either easier or more difficult to diagnose, the sensitivity and specificity can actually appear to be prevalence-dependent.

There are a few other test performance characteristics that are frequently reported and worth defining here. Accuracy is the overall probability that a test provides the right diagnosis, combining results for both truly positive and truly negative subjects. From Table 3.1, we can estimate it as $P(T+|D+) + P(T-|D-) = (a+d)/(a+b+c+d) = 31/70 + 33/70 = 0.91$. Like the PPV and NPV, however, accuracy is also influenced by the prevalence of the disease, so is not a property of the diagnostic test alone. Precision differs from accuracy in that precision refers to the consistency of the test if one were to repeat it over and over on the same patient, under the same conditions. It is closely related to intraobserver reliability which is the chance that the same interpreter of a given test, when presented with it again at different time, will come to the same conclusion. This is more an issue for tests that are subject to clinical interpretation, such as X-rays, and less an issue for numerical values from blood tests, for example. Interobserver reliability relates to the chance that two different health-care workers interpreting the same test in the same patient will come to the same conclusion, in terms of classifying the test result as positive or negative. It is frequently reported as a κ statistic (defined as the agreement observed beyond chance divided by the agreement possible beyond chance), where 1.0 is perfect agreement, 0 is no more agreement than would be expected by chance alone, and -1.0 is perfect disagreement. In general, a $\kappa < 0.4$ reflects poor agreement, 0.4-0.6 is fair, 0.6-0.8 is good, and > 0.8 is excellent.[7] For example, at the McGill University Health Sciences Center, the interobserver agreement among radiologists for the diagnosis of choledocholithiasis on magnetic resonance cholangiopancreatography (MRCP) is 0.82, which is very good agreement.[8] Lastly, utility is often used in the diagnostic test context as a somewhat vague measure that refers to the clinical usefulness of a diagnostic test. A test like abdominal ultrasound, which accurately identifies gallbladder stones in completely asymptomatic patients may have excellent sensitivity and specificity, but may have poor utility if the results do not help modify your management or your approach to the patient.

Odds and Likelihood Ratios

Before discussing odds ratios and their relationship to likelihood ratios, we will first need to discuss the concept of pretest (or prior) and posttest (or posterior) probabilities of disease. Pretest probability is, as its name implies, the chance that one would give the patient of having the disease before knowing the results of the test in question. The posttest probability is the updated chance one would give them of having the disease after one has found whether the test result is positive or negative. In many cases, the pretest probability is estimated by the prevalence of disease in the population of interest. In other situations, it may be the posttest probability of a prior diagnostic test. For example, if we consider the clinical examination as a diagnostic test, with the prevalence of disease being the pretest probability of disease, then the posterior probability of having the disease, following the history and physical exam, becomes the pretest probability for the next investigation (e.g., bloodwork, X-ray, etc.).

For example, after hearing about a 65 year old patient with rectal bleeding, one might initially estimate the probability of cancer at 20%. After your history reveals weight loss and a feeling of incomplete evacuation, your posttest probability might rise to perhaps 80%. This 80% then becomes the pretest probability for your next intervention which might be a digital rectal exam. The uncertain impression on the rectal exam of a vague mass then further increases that posttest probability to, say, 95%. While exact probabilities may be difficult to assess even after a careful literature review, the clinician will always keep some idea of the probabilities in mind when deciding on a diagnosis for any patient.

Let us consider another more realistic example from the literature. In a study looking at the detection of splenomegaly by physical exam, 118 patients were examined using a variety of maneuvers, including splenic percussion, Traube's space percussion, supine and right lateral decubitus palpation, and Middleton's maneuver.[9] Sensitivities and specificities were calculated for the individual maneuvers as well as the conditional sensitivity (74%) and specificity (90%) for palpation (positive defined as spleen probably palpable or definitely palpable), given a positive Traube's space percussion (defined as uncertain, probably dull or definitely dull). No palpation maneuver was found to be better than any other. Let us say one is rounding on a ward in which it is estimated that the prevalence of splenomegaly is 10%. The pretest probability is then, by default, also 10% before physical examination. With a sensitivity of 62% and specificity of 72% for Traube's space percussion, the posttest probability of splenomegaly rises to 19.7% with positive percussion. This then becomes the pretest probability before palpation. If one then palpates for a spleen, and thinks the spleen is probably palpable, the posttest probability rises to 64.5%. This is now the pretest probability before abdominal ultrasound, and so on. The calculation of these probabilities is discussed below.

The odds of having a certain disease are simply defined as the probability of a patient having the disease divided by the probability of the patient not having it, or P(D+) / (1-P(D+)). Therefore, if the probability of a gastric ulcer being malignant based on its large size at endoscopy is 67%, then the odds of it being malignant are 0.67/(1-0.67), or 2 to 1.

Likelihood ratios are a form of odds ratio. The LR for a positive test (LR+) represents the odds of a person having a disease if the test is positive, by comparing the probability of a diseased person having a positive test with that of a person not having the disease getting that same result, and is defined as...

$$\text{LR+} = \frac{P(T+|D+)}{P(T+|D-)} = \frac{\text{True Positive Rate}}{\text{False positive Rate}} = \frac{\text{Sensitivity}}{1 - \text{Specificity}}$$

Referring to Table 3.1, we can estimate the LR+ by...

$$\frac{a/(a+c)}{b/(b+d)}$$

In contrast, the LR of a negative test (LR-) is the odds of a person not having the disease if the test is negative and is represented by the following...

$$LR- = \frac{P(T-|D-)}{P(T-|D+)} = \frac{\text{True Negative Rate}}{\text{False negative Rate}} = \frac{1 - \text{Sensitivity}}{\text{Specificity}}$$

Referring to Table 3.1, we can estimate the LR- by...

$$\frac{c \,/\, (a+c)}{d \,/\, (b+d)}$$

Likelihood ratios (LR) are helpful in comparing various tests because they integrate the sensitivity and the specificity and tell you, at a glance, how discriminate the test is and what its best role is, i.e., whether the test is better for ruling in or out disease. They are also independent of disease prevalence, although the caveats relating to this point for sensitivities and specificities are of course inherited by LRs.

Specificity-corrected sensitivity (SC-sensitivity) and sensitivity-corrected specificity (SC-specificity) are two other related statistics that have been described to attempt to similarly capture the balance between sensitivity and specificity.[10] The former is defined as the sensitivity divided by "lack of specificity" or sensitivity(1-specificity) which is the same as a LR+. The latter corrects specificity for "lack of sensitivity" or specificity/(1-sensitivity) and is the same as the reciprocal of a LR-. This terminology, however, is not commonly used and the terms LR+ and LR- are generally preferred.

A diagnostic test exhibiting a LR of 1 does not provide any useful information. It means that people with and without disease have an equal chance of having a positive test. A coin toss, if it were used as a diagnostic test, would have a LR of 1; the chance of a diseased person getting "heads" is equal to that of a person without the disease. If a test displays a LR+ of far less than one, it is, in general, helpful at ruling out disease and one that is much greater than one is, in general, helpful at ruling in disease (although this does depend on the pretest odds of disease).

LRs do not apply just to dichotomized test results, but can also be calculated for individual ranges of values of continuous test results. Unfortunately, there is a paucity of studies that report sufficient data to allow for their accurate estimation. Their importance, however, lies in the fact that not all values below or above any cut off for a continuous diagnostic test have the same meaning. Consider, for example, the use of body temperature in a patient with acute abdominal pain to help diagnose acute appendicitis. One is aware that anything above 37.7°C is abnormal. However, one intuitively accepts that the higher the temperature, the more likely appendicitis may be. A value of 37.9°C is not the same as 38.9°C in its ability to predict appendicitis. This is where LRs for continuous tests are useful. LRs allow one to pick selected ranges of a test value and examine how helpful the individual ranges of values are. The LR for a range is the probability of having the disease given that one's test result falls in that range divided by the probability of not having the disease given one's result falls in that same range. In fact, when different clinical and laboratory values were recently studied to determine their usefulness in diagnosing acute appendicitis, the likelihood ratios for ranges into which these two temperatures fell were found to be 1.16 and 5.59, respectively.[11] That is, the former is not very helpful in altering your pretest suspicion of appendicitis, whereas the second will increase it markedly (even though both were above the upper limit of normal).

As another example, consider the performance of various biochemical markers for the diagnosis of acute pancreatitis. The data in Table 3.3 were derived from scatter plots (see Fig. 3.3, discussed more fully later in this Chapter).[12] The first step towards estimating the sensitivity and specificity of a given cutoff is to construct a 2x2 table. Table 3.4 is such a table, using the previously recommended cutoff for the assay of 80 U/L.[13] The sensitivity estimate is therefore 31/37 = 0.84 and the specificity estimated at 98/99 = 0.9898 at this cutoff. By substituting these into the above equations, the LR is 0.84/(1-0.9898) = 82. This means a lipase > 80 U/L is extremely helpful at confirming pancreatitis, although the background prevalence must also be considered before any final diagnosis is made.

But how about intermediate lipase values? How should a value of 70 U/L be interpreted? How about a value of 20 U/L? To calculate the LR for any given range we simply assume that the range 40.1-60 U/L, for example, is considered a positive test and then create a corresponding 2x2 table (Table 3.5), and calculate the resulting sensitivity and specificity. As it happens, there is a shortcut using Table 3.3: take the fraction of patients with pancreatitis who had lipase values within that range (the true positive rate) and divide it by the fraction of patients without pancreatitis who also had lipase values within that range (the false positive rate). That is, 3/37 divided by 19/98 = 0.42. The other LRs are displayed in Table 3.3 across from their respective ranges. They demonstrate that lipase levels less than 60 U/L are helpful at ruling out pancreatitis (with those less than 40 being the most helpful), levels above 80 U/L are helpful at ruling in pancreatitis, and levels between 60 and 80 U/L do not substantially alter your diagnostic certainty above and beyond your pretest clinical suspicion.

There are many other situations where likelihood ratios for different ranges of values of a continuous variable would be helpful. Serum alpha-fetoprotein for diagnosing hepatocellular carcinoma is one. Common bile duct diameter for predicting choledocholithiasis is another. One important continuous variable, for which there is likelihood ratio data, is ferritin used to diagnose iron deficiency anemia (Table 3.6). Because in addition to being a measure of total iron stores, it is an acute phase reactant, a normal ferritin level is difficult to define, especially in the elderly who may have chronic comorbidities. Therefore, one study looked at geriatric patients with anemia and determined iron status in many of them by bone marrow examination. The results and likelihood ratios are summarized in Table 3.6.[14] One can see that despite its limitations, a ferritin above 100 mg/L virtually rules out iron deficiency. Somewhat disturbingly, however, even a ferritin of 30 mg/L in this population, which is above the lower limit of normal in most centers, still increases your pretest odds of disease three-fold and is likely to warrant further investigation, such as a colonoscopy. Thus, having likelihood ratios for these continuous variables would allow clinicians to come to much more informed conclusions than using an often arbitrary cutoff suggested by a test's manufacturer. As always, one must consider to what degree the test characteristics reported in the literature apply to specific individuals seen in any clinic.

Bayes Theorem and Posttest Probabilities

To understand how the likelihood of disease evolves when more than one test is performed, one needs to first understand a few properties about probabilities. Suppose

3

Table 3.3. Lipase as a marker for pancreatitis (Data from Steinberg et al)[12]

Range of lipase (U/L)	Acute pancreatitis Present	Absent	Totals	Likelihood Ratios
0-40	2	76	78	0.07
40.1-60	3	19	22	0.42
60.1-80	1	2	3	1.3
> 80	31	1	32	82
TOTALS	37	98	135	

Table 3.4. Acute pancreatitis (Data from Steinberg et al)

		D+	D-	
Lipase > 80 U/L	T+	31	1	32
	T-	6	97	103
		37	98	135

Table 3.5. Acute pancreatitis (Data from Steinberg et al)

		D+	D-	
Lipase 40.1-60 U/L	T+	3	19	22
	T-	34	79	113
		37	98	135

Table 3.6. Ferritin levels in geriatric patients with anemia (Data from Guyatt et al)

Ferritin Range	Iron Deficiency Anemia		Likelihood Ratio	Role of Values in that Range in Predicting Iron Deficiency
	PRESENT	ABSENT		
≤ 18 μg/L	47	2	42	Strongly Rule In
18.1 to 45 μg/L	23	13	3.1	Weakly Rule In
45.1 to 100 μg/L	7	27	0.5	Weakly Rule Out
>100 μg/L	8	108	0.1	Strongly Rule Out

two diagnostic tests are performed on a given patient. Let us call the event that test one is positive A, and the event that the second test is positive B. The probability that both tests are positive or the joint probability that both events A and B occur is a simple calculation if the two tests are independent (see the section on the rules of probability in Chapter 2). It is simply the product of the probabilities of each event: P(A and B) = P(A) x P(B). However, if the two test results are not independent, then the probability that the second test is positive depends on whether the first test is positive. Thus to calculate the probability that both tests are positive, we first calculate the probability that the first test is positive, and then multiply this by the probability that the second test is positive given that the first test is positive. That is, P(A and B) = P(A) x P(B|A). The calculation of this quantity is easier with independent tests because, if the tests are independent, then P(B) = P (B|A). This is because the probability of B is the same whatever the outcome of the first test, assuming independence between the tests.

Bayes Theorem is a general theorem, directly derivable from the basic rules of probabilities, that involves conditional probabilities. In the case of diagnostic tests, it is useful because it displays the relationship between sensitivity, specificity, prevalence and predictive values. It states that the P(D+|T+) or PPV is equal to...

$$\text{PPV} = \frac{P(T+|D+) \times P(D+)}{P(T+|D+) \times P(D+) + P(T+|D-) \times P(D-)}$$

More simply, by substituting familiar terms, ...

$$\text{PPV} = \frac{\text{Sensitivity} \times \text{Prevalence}}{\text{Sensitivity} \times \text{Prevalence} + (1\text{-Specificity}) \times (1\text{- Prevalence})}$$

Similarly, the formula for NPV or P(D-|T-) can be derived...

$$\text{NPV} = \frac{\text{Specificity} \times (1 - \text{Prevalence})}{\text{Specificity} \times (1 - \text{Prevalence}) + (1 - \text{Sensitivity}) \times (\text{Prevalence})}$$

Bayes Theorem can be used to quickly calculate predictive values, if sensitivity and specificity are known, without first presenting the data in a 2x2 table format. For example, in the case of a patient with a pancreatic mass, let us say that the pretest probability of adenocarcinoma is 90%. If the patient undergoes a fine needle aspirate, which, let us say, in your center has a sensitivity of 60% and a specificity of 98%, what is the predictive value of a negative test (NPV)? We calculate: 0.98 x 0.10 / (0.98 x 0.10 + 0.40 x 0.90) = 0.18 or 18%. This is very poor but not altogether unexpected. As we said above, even a test with reasonable sensitivity, given a high prevalence or pretest likelihood of disease, can have a low NPV. The pretest probability of disease was so high that a negative test influences your suspicion of disease very little. The PPV, in contrast, is 0.60 x 0.90 / (0.60 x 0.90 + 0.02 x 0.10) = 0.996 or 99.6%. This illustrates that when the pretest probability and specificity are both high, as is the case here, the PPV is usually very high. In fact, if you had performed a test that was less specific, like a CT scan of the abdomen, the PPV would still be quite high because of the high pretest probability. For example, substituting a lower specificity of 70%, the PPV remains high at 95%.

LRs can also be used to obtain a posttest probability of disease. One first needs to know the prevalence of disease or the pretest probability of disease based on your clinical exam, for example. You then convert that probability to pretest odds of disease as described above. The LR multiplied by the odds then gives a posttest odds of disease.

$$\text{Posttest odds} = \text{pretest odds} \times \text{LR}$$

The resulting odds can be converted back to a probability using the formula,

$$\text{Probability} = \frac{\text{odds}}{(\text{odds} + 1)}$$

For example, Detsky et al,[15] have published a multifactorial index for the prediction of cardiac complications after noncardiac surgery. From that paper, the pretest probability of a cardiac complication after abdominal surgery is quoted as 8.0% (95% CI 3.4-13.6%). If your patient has had a previous remote MI, is over 70 years old, has a Canadian Cardiovascular Society Class of 3, and is going for an emergency operation, her LR for a cardiac event is 7.54. They provide a nomogram that allows you to take a straight edge, line it up with the pretest probability and the LR, and read off the posttest probability. From the nomogram, the posttest probability looks to be about 40% but, as an exercise, let us calculate it and check. Her pretest odds are 0.08/(1-0.08) = 0.08 / 0.92 or 0.087. Multiplying this by the LR, we get 0.087 x 7.54 = 0.656. Then converting this back to probability, 0.656 / (1 + 0.656) = 0.396 or 39.6% which is very close to the estimate from the nomogram. Nomograms are an accurate and quick way of converting pretest to posttest probabilities using LRs, without a calculator, and can be carried around in a pocket. Figure 3.2 is one such nomogram.

When interpreting a series of diagnostic tests, for example, an ultrasound followed by a HIDA scan for diagnosing acute cholecystitis, assuming the test results are all independent of one another, all you have to do is multiply the posttest odds of the first test by the likelihood ratio of the second test to get the second posttest odds, and so on. If the tests are not independent, however, then the likelihood ratio of the second test must be one that takes into account that the first test was positive.

$$\text{Odds of disease for positive tests A and B} = \text{pretest odds} \times \text{LR(A)} \times \text{LR(B|A)}$$

It is very often the case that diagnostic test results from two or more tests are not independent. For example, tests for a given parasite may tend to be correlated, since a patient with a high degree of infection may tend to test positively on two tests, while low degrees of infection may tend to be missed by all tests, thus inducing dependence. Therefore, posttest probabilities from two or more tests must be interpreted with great caution.

Choosing the Best Cutoff Value

In preliminary studies of a continuous diagnostic test, one of the goals is to decide on the best discrimination level or "cutoff" value to recommend to be used in subsequent work. As we discussed above, cutoffs that improve sensitivity frequently lower specificity and vice versa, therefore one needs a cutoff that is "balanced" in terms of the performance it portends in these two areas. The problem is disease-specific, since an ideal tradeoff between the false positive and false negative

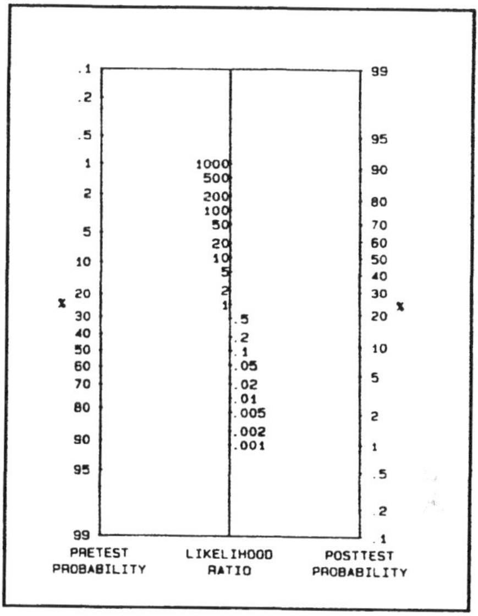

Fig. 3.2. Nomogram for the calculation of posttest probability of disease. A straight-edge through the pretest probability of disease or prevalence and the likelihood ratio indicates the posttest probability of a cardiac complication after noncardiac surgery. (Reproduced with permission from Detsky et al.).

rate for a relatively benign condition may be highly inappropriate for a more serious condition. We have already discussed above how one can individually calculate sensitivities and specificities at different cutoffs, and how one can determine likelihood ratios for different ranges of a continuous variable, and thereby, pick the ideal cutoff. There are other methods that can be used to pick that ideal cutoff and these will be discussed below.

Figure 3.1 shows two overlapping distributions, one of the test results from a diseased population and one of results from a nondiseased population, over a range of values for a hypothetical diagnostic test. It is unfortunately usual for the two populations that one wants to discriminate to overlap in this way, to a greater or lesser degree. One can see that as one moves the cutoff value from left to right, one picks up more diseased people (increases sensitivity) until one gets to a point where one is now misclassifying many nondiseased individuals as diseased (decreasing specificity). Similarly, attempts at moving the test cutoff leftward so that one avoids misclassifying nondiseased subjects (increases specificity), eventually cause the test to miss diseased individuals (decrease sensitivity). This type of graphical representation of sensitivity simultaneously with specificity allows one to visually pick a cutoff that confers the minimum overlap, and therefore, balances sensitivity and specificity.

The use of scatter plots is another related qualitative method. A scatter plot displays the individual test results for each patient with the disease and plots those values in a column next to a similar display of the results obtained from patients with a different disease or those with no disease. Each dot represents the results from one patient, and at a glance one can use a horizontal straight edge to see which cutoff value will leave most dots in the disease column above the line and most control dots below the line. Figure 3.3 is a scatter plot of the lipase data from the acute pancreatitis study, referred to above, and the horizontal line represents a reasonable cutoff value. Comparisons were made to acute appendicitis, biliary tract disease, gynecologic disease, large bowel disease, peptic ulcer disease, perforated viscous, and small bowel disease.

Another way to determine the ideal cutoff value for a test is with a Receiver Operating Characteristic curve or ROC curve. An ROC curve is simply a graph that plots true positive rates against false positive rates for a series of cutoff values, or in other words, sensitivity is plotted on the Y-axis versus [1–specificity] on the X-axis for each cutoff value. An ideal cutoff might give the test the highest possible sensitivity with the lowest possible false positive rate (i.e., highest specificity). This is the point lying geometrically closest to the top-left corner of the graph (where the ideal cutoff value with 100% sensitivity and specificity would be plotted), although of course disease-specific considerations may suggest other more clinically useful compromises between sensitivity and specificity. To illustrate this, we will look at a study seeking predictors of common bile duct (CBD) stones in patients undergoing laparoscopic cholecystectomy.[16] In this study, historical, biochemical, and ultrasonographic data were collected on patients who underwent endoscopic retrograde cholangiopancreatography (ERCP) and laparoscopic cholecystectomy. ROC curves were then constructed for the various biochemical tests to choose the best cutoffs for the prediction of CBD stones. Figure 3.4 illustrates the ROC curve for alkaline phosphatase. The plotted value which was geometrically closest to the upper left corner of the graph was 300 U/L, and so this was taken to be the ideal cutoff.

In certain situations, however, where for example, missing the diagnosis would be devastating (e.g., appendicitis), or treating a normal person by mistake would be particularly dangerous (e.g., Whipple's procedure for a benign pancreatic mass), the top-left corner may not be the ideal point. In the former, one would be willing to sacrifice specificity for a very sensitive test, and for the latter, a test with high specificity would be important. Therefore, picking the ideal cutoff is, to some extent, dependent on the clinical context.

Another use of the ROC curve is to qualitatively and quantitatively assess the discriminating ability of a test. A cutoff whose true positive rate equals its false positive rate will have a likelihood ratio of one. If all the possible cutoffs of the test have a LR of one, the test would then graphically look like the dotted line (at 45° to the axes) in Figure 3.4, which represents such a nondiscriminant test. The area under an ROC curve can be used as an overall assessment of its discriminating ability.[17,18] A nondiscriminant test has an area under the curve of 0.5. One can see that alkaline phosphatase is not a particularly discriminating test when it comes to predicting CBD stones before laparoscopic cholecystectomy. While a discussion of the formulae for quantitative ROC curve analysis is beyond the scope of this Chapter, user-friendly software is available.

Another interpretation of the area under the ROC curve is as follows. Suppose one truly diseased subject and one truly nondiseased subject are selected at random from the populations of all diseased and all nondiseased subjects, respectively. It can

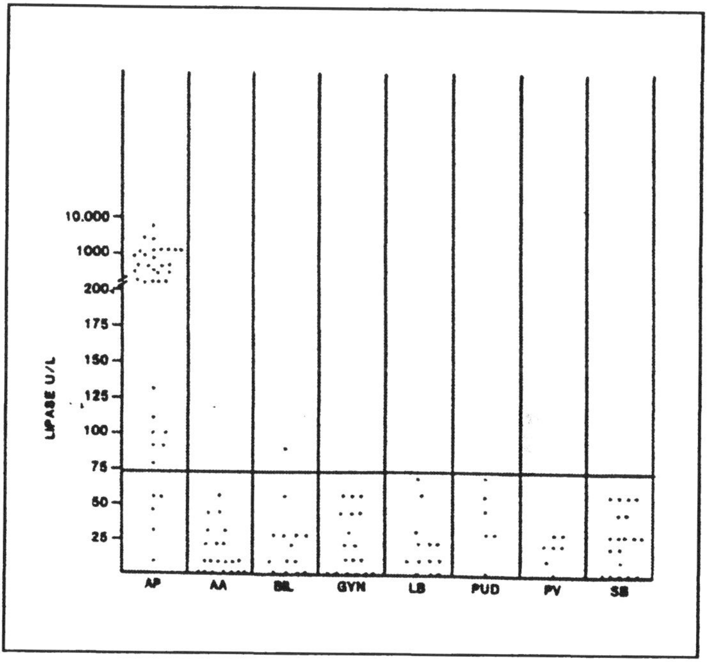

Fig. 3.3. Scatterplot of lipase results for 8 categories of disease representing the ultimate diagnoses in patients presenting with acute abdominal pain: Acute Pancreatitis (AP); Acute Appendicitis (AA); Biliary Tract Disease (BIL); Gynecologic Disease (GYN); Large Bowel Disease (LB); Peptic Ulcer Disease (PUD); Perforated Viscus (PV); Small Bowel Disease (SB). The "best cutoff" recommended by Steinberg et al is represented by a horizontal line. (Reproduced with permission from Steinberg et al).

be proven mathematically that if each of these subjects are given the test, then the probability that the test correctly classifies these subjects (i.e., the diseased subject scores "higher" than the nondiseased subject) is exactly equal to the area under the ROC curve.

The more mathematically inclined of you may also notice that the slope of the ROC curve at any given point (the true positive rate over the false positive rate) is in fact the likelihood ratio of a positive test that uses that point as its cutoff value. Therefore, the slope of a line drawn at a tangent to any point on the ROC curve yields the likelihood ratio for that cutoff. The steeper the tangent, the more that cutoff is useful at ruling in disease; the less steep it is, the more that cutoff is useful at ruling out disease. The ideal cutoff value usually has a tangent with a slope of near 1 because a test result that lands right on the cutoff (a borderline result) will usually be of little use in discriminating diseased from nondiseased subjects. The slope of

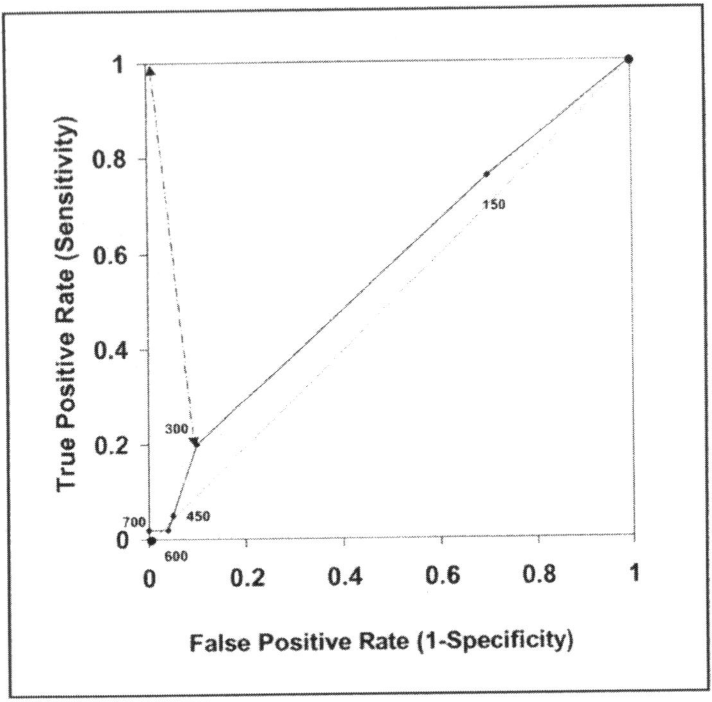

Fig. 3.4. This ROC curve displays various alkaline phosphatase cutoff values (U/L) for the prediction of common bile duct stones in patients undergoing laparoscopic cholecystectomy. The value geometrically closest to the top-left corner (100% sensitivity and specificity) of the graph was 300 U/L. The dashed arrow illustrates the shortest distance. The dotted 45° line represents a completely nondiscriminant test (Reproduced with permission from Barkun et al).

the nondiscriminant 45° line is 1 at all points, and a test whose every cutoff value has a likelihood ratio of one has, by definition, no discriminating ability.

The Screening Test—Special Considerations

The decision of whether or not to use a test to screen for a disease involves several issues.[19] Is the disease a significant enough health problem that it merits the cost of a screening program? Is the prognosis of the disease, as it is diagnosed currently, poor and in need of improvement? If the disease were picked up at an earlier stage, would the treatment then be more effective? Finally, will people, and their primary physicians, comply with the screening protocol, i.e., is it safe, convenient, etc.? As well, after a screening program is implemented, "healthy" people may be labeled with a "pre-disease" which often has its own behavioral implications, as was illustrated by the increased absenteeism seen when steel workers had their hypertension identified and treated.[20]

In their early development, the attempts to demonstrate the validity of these programs may include proving improved prognosis of a group identified with the screening test compared to a control group that is not screened. Although a full discussion of this topic[21] is beyond the scope of this Chapter, one should be aware of the potential for bias here because of failure to correct for the "lead time" brought about by early detection, also called the "zero-time shift". Lead time bias is the extra survival time or disease-free survival time that you get from simply diagnosing a condition earlier in the screened cohort compared to the controls, especially when historical controls are used. For example, comparing the prognosis of an asymptomatic mammographic abnormality is very different from a palpable breast lump.[22] As well, "volunteers" in screening studies are almost always healthier in other ways than the general population. Because of all of these effects, even when the screening test is in fact useless, early diagnosis may appear to improve survival if one does not correctly adjust for all biases.

The ideal screening test is highly sensitive. This is because subsequent tests will not be arranged unless this test is positive, and so false negatives will exclude patients from further investigation, in general. The test should be followed by a more specific confirmatory test, if the original test lacks specificity, if the treatment is hazardous or the disease label is devastating (as with HIV testing). Remember also that since these tests are generally performed in populations with a low prevalence of the disease, the PPV will be low even if the test is reasonably specific. For this reason, seeking out a target screening group that is "at high risk" for the given condition is often desirable. Even then, however, the prevalence of disease is usually not very high.

Screening tests that are intended to be repeated, for example, on an annual basis, may sometimes be reported with two sets of performance statistics. The first is one that assumes that the test is done once, on its own. The second, is the sensitivity of two or more years of repeated testing. Even an insensitive test, if repeated, could potentially achieve a high "program sensitivity". The studies on fecal occult blood testing used to screen for colorectal cancer exemplify this point. Here, in addition to the mathematical reasons for overall sensitivity increasing with the number of tests, repeated testing has another advantage in that it involves multiple opportunities to catch an intermittently bleeding polyp or cancer. As well, about half of the cancers discovered in three large clinical trials[23-25] were in fact not discovered by the screening test (fecal occult blood test) per se, but rather presented clinically in the time between annual tests.[26] However, because the cancers were detected while patients were enrolled in the screening program, they contribute to the program sensitivity. When a simulation model was prepared for an American panel charged with the task of developing clinical guidelines for colorectal cancer screening, they assumed a sensitivity of doing a single occult blood test to be 60% based on the available literature.[27] With this single-test sensitivity, the overall program sensitivity was 93% in uncovering colorectal cancer. Which is the real sensitivity? It depends on whether you are looking for the sensitivity of the test itself or of the screening program. Both are useful, as long as one keeps in mind that the implications of a single occult blood positive test are determined by its sensitivity on its own and not the sensitivity quoted in a study examining the performance of an entire screening program. The same is true about specificity, NPV and PPV.

Lastly, when one is trying to see whether a screening test is helpful, one must specifically look out for cointervention bias in the study that compares the outcome of patients who have had the screening test plus an intervention with that of unscreened patients. This bias occurs because patients in screening programs are treated differently, have closer follow-up, more opportunity to ask questions, etc. than their nonscreened counterparts, which has an effect on their health or outcome quite separate from either the success of the screening test or its associated intervention. Again, examining a study on interventions or therapy is beyond the scope of this Chapter, but for more information about bias in a study on therapy, we refer you to Guyatt et al.[28,29]

Approach to a Study Examining a Diagnostic Test

Reading papers reporting on the properties of diagnostic tests can be confusing. The quality of the study methodology varies widely and the relevance of the test to clinical practice is not always mirrored by the conclusions of the authors. In general, one needs to extract from the study who it was that was being tested, exactly how the test was done, what other test (or "gold standard") it was being compared to, and finally, how this new test fared in comparison. Even if the test is appropriate and helpful, practical issues such as cost, availability, safety, and reliability will also have to enter into the analysis. Lastly, one needs to know if the condition is appropriate to diagnose in the first place, and whether the patient will be better off for having had the test. Is it treatable at the stage at which it is being diagnosed? Will the patients agree to have it done? Will it replace an existing test or will this be the third or fourth invasive examination for that patient? The following questionnaire, formulated by the Evidence Based Medicine Working Group and published in JAMA,[30,31] incorporates these concerns. It is a useful checklist to go through when evaluating a report of a new diagnostic test:

1. Has there been an independent, "blind" comparison with a "gold standard" of diagnosis?
2. Has the diagnostic test been evaluated in a patient sample that included an appropriate spectrum of mild and severe, treated and untreated, disease, plus individuals with different but commonly confused disorders?
3. Was the setting for this evaluation, as well as the filter through which study patients passed, adequately described?
4. Have the reproducibility of the test result (precision) and its interpretation (observer variation) been determined?
5. Has the term normal been defined sensibly as it applies to this test?
6. If the test is advocated as part of a cluster or sequence of tests, has its individual contribution to the overall validity of the cluster or sequence been determined?
7. Have the tactics for carrying out the test been described in sufficient detail to permit their exact replication?
8. Has the utility of the test been determined? That is, is the patient better off for having the test? Has a treatable disorder been identified? Are other tests still necessary?

Summary

The thoughtful interpretation of diagnostic tests, whether they are in the form of historical points, physical exam maneuvers, laboratory measurements, radiologic procedures, cytology or biopsy, etc., is essential to planning appropriate intervention and/or accurately predicting prognosis. In this Chapter, several test performance characteristics were outlined that help evaluate the usefulness of a test. We demonstrated how to estimate these characteristics from a 2x2 table. Sensitivity is the chance of test being positive in disease (PiD) and specificity is the chance of being negative in health (NiH). It was explained that sensitivity and specificity are inherent to a test, whereas PPV an NPV are prevalence-dependent. Spectrum bias can cause sensitivity and specificity to also appear dependent on prevalence in certain situations. Tests with high sensitivities or NPV's rule out disease and high specificities and PPV's rule in disease, which can be remembered with the mnemonics, SnNout and SpPin, respectively. Likelihood ratios, the ratio between the true positive rate and the false positive rate, are a prevalence-independent way of comparing the performance of diagnostic tests; and can allow one to break down continuous variables into ranges of values so that one need not rely solely on one individual cutoff value. One can also use LRs to convert pretest odds, which are calculated from pretest probability or prevalence, to posttest odds and then back again to a posttest probability of disease, with or without the help of a nomogram. Bayes Theorem can be used to calculate predictive values from sensitivity, specificity and prevalence. Different ways to pick an ideal cutoff for a diagnostic test were covered including visualization of the overlap of normal and diseased population distributions, likelihood ratios, scatter plots, and ROC curves. In the latter, the true positive rate is plotted against the false positive rate and the point closest to the top-left of the graph often represents the best balance between sensitivity and specificity. The area under an ROC curve represents the discriminating ability of a test. Special considerations for the evaluation of a screening test were discussed, touching on the concepts of the ideal test, program sensitivity, the zero time shift phenomenon (or lead time bias), and cointervention bias. Lastly, the questionnaire published by the Evidence Based Medicine Working Group was presented, which gives a step-by-step approach to determine the validity and applicability of a study examining a diagnostic test.

While many issues have been covered, one book Chapter can never detail every issue of the vast literature on this topic. The sensitivity, specificity, PPV, NPV, and prevalence estimated from 2x2 table data are examples of binomial variables, so that 95% confidence intervals can be calculated from formulae given in Chapter 2. Similarly, properties from two different tests can be compared using techniques for comparing two proportions. Confidence intervals for likelihood ratios[32] and ROC curves[33] and the areas under them are described elsewhere.

The statistical analysis of diagnostic test data is still a very hot research topic. Methods on the appropriate analysis of correlated tests[34] and on the estimation of test characteristics when an imperfect "gold standard" is used[35,36] are examples of areas where further research is required.

Nevertheless, an understanding of the material in this Chapter should serve to prepare the clinician to be an aware user of diagnostic tests.

3

Selected Readings

1. Vignault F, Filiatrault D, Brandt ML et al. Acute appendicitis in children: Evaluation with US. Radiology 1990; 176: 501-4.
2. Sackett DL, Haynes RB, Guyatt GH et al. The Interpretation of diagnostic data. In: Clinical Epidemiology: A Basic Science for Clinical Medicine. 2nd ed. Toronto, Canada: Little, Brown and Company; 1991:69-152.
3. Verilli D, Welch HG. The impact of diagnostic testing on therapeutic interventions. JAMA 1996; 275: 1189-91.
4. Lachs MS, Nachamkin I, Edelstein PH et al. Spectrum bias in the evaluation of diagnostic tests: lessons from the rapid dipstick test for urinary tract infection. Ann Intern Med 1992; 117:135-40.
5. Ransahoff DF, Feinstein AR. Problems of spectrum bias in evaluating the efficacy of diagnostic tests. N Engl J Med 1978; 299:926-30.
6. Shapiro MF, Lehman AF, Greenfield S. Biases in the diagnosis of depression. Arch Intern Med 1983; 143:2085-8.
7. Schoenfeld P, Guyatt G, Hamiton G et al. An evidence-based approach tô gastroenterology diagnosis. Gastroenterology 1999; 116:1230-37.
8. Reinhold C, Taourel P, Bret PM et al. Choledocholithiasis: Evaluation of MR cholangiography for diagnosis. Radiology 1998; 209:435-42.
9. Barkun AN, Camus M, Green L et al. The bedside assessment of splenic enlargement. Am J Med 1991; 91:512-8.
10. Tebaarwerk GJM. Measuring the efficacy and cost-effectiveness of laboratory tests. Annals RCPSC 1995; 28:217-220.
11. Andersson RE, Hugander AP, Ghazi SH et al. Diagnostic value of disease history, clinical presentation, and inflammatory parameters of appendicitis. World J Surg 1999; 23:133-140.
12. Steinberg WM, Goldstein SS, Davis ND et al. Diagnostic assays in acute pancreatitis. Annals Int Med 1985; 102:576-80.
13. Verduin PA, Punt JMH, Kreutzer H. Studies on the determination of lipase activity. Clin Chim Acta 1973; 55:269-89.
14. Guyatt G, Patterson C, Ali M et al. Diagnosis of iron deficiency anemia in the elderly. Am J Med 1990; 88:205-9.
15. Detsky AS,. Abrams HB, Forbath N et al. Cardiac assessment for patients undergoing noncardiac surgery. Arch Int Med 1986; 146:2131-4.
16. Barkun AN, Barkun JS, Fried GM et al. Useful predictors of bile duct stones in patients undergoing laparoscopic cholecystectomy. Ann Surg 1994;220:32-9.
17. Metz CE, Kronman HB. Statistical significance of tests for binomial ROC curves. J Math Psychol 1980; 22:218-43.
18. Dorfman DD, Alf E Jr. Maximum-likelihood estimation of parameters of signal-detection theory and determination of confidence intervals-rating-method data. J Math Psychol 1969; 6:487-96.
19. Sackett DL, Haynes RB, Guyatt GH et al. Early diagnosis. In: Clinical Epidemiology: A Basic Science for Clinical Medicine. 2nd ed. Toronto, Canada: Little, Brown and Company; 1991:153-170.
20. Taylor DW, Haynes RB, Sackett DL et al. Long-term follow-up of absenteeism among working men following the detection and treatment of their hypertension. Clin Invest Med 1981; 4: 173.
21. Laupacis A, Wells G, Richardson WS et al. Users' guides to the medical literature. V. How to use an article about prognosis. Evidence-Based Medicine Working Group. JAMA 1994. 272:234-7.
22. Shapiro S. Evidence of screening for breast cancer from a randomized trial. Cancer (suppl) 1977; 39:2772

23. Mandel JS, Bond JH, Church TR et al. Reducing mortality from colorectal cancer by screening for fecal occult blood. N Engl J Med 1993; 328:1365-71.
24. Hardcastle JD, Chamberlain JO, Robinson MHE et al. Randomized controlled trial of faecal-occult-blood screening for colorectal cancer. Lancet 1996; 348:1472-7.
25. Kronborg O, Fenger C, Olsen J et al. Randomised study of screening for colorectal cancer with faecal-occult-blood test. Lancet 1996;348:1467-71.
26. Simon JB, Fletcher RH. Should all people over the age of 50 have regular fecal occult-blood tests? [Clinical Debate] N Engl J Med 1998; 338(16):1151-1155.
27. Winawer SJ, Fletcher RH, Miller L et al. Colorectal cancer screening: clinical guidelines and rationale. Gastroenterology 1997; 112: 594-642.
28. Guyatt GH, Sackett DL, Cook DJ. Users' guides to the medical literature. II. How to use an article about therapy or prevention. A. Are the results of the study valid? Evidence-Based Medicine Working Group. JAMA 1993; 270:2598-601.
29. Guyatt GH, Sackett DL, Cook DJ. Users' guides to the medical literature. II. How to use an article about therapy or prevention. B. What were the results and will they help me in caring for my patients? Evidence-Based Medicine Working Group. JAMA 1994; 271:59-63.
30. Jaeschke R, Guyatt G, Sackett DL. Users' guides to the medical literature. III. How to use an article about a diagnostic test. A. Are the results of the study valid? Evidence-Based Medicine Working Group. JAMA 1994; 271:389-91.
31. Jaeschke R, Guyatt GH, Sackett DL. Users' guides to the medical literature. III. How to use an article about a diagnostic test. B. What are the results and will they help me in caring for my patients? The Evidence-Based Medicine Working Group. JAMA 1994; 271:703-7.
32. Centor RM. Estimating confidence intervals of likelihood ratios. Medical Decision Making 1992; 12:229-233.
33. Metz C. Basic principles of ROC analysis. Seminars in Nuclear Medicine 1978; 8(4):283-298.
34. Qu YS, Hadgu A. A model for evaluating sensitivity and specificity for correlated diagnostic tests in efficacy studies with an imperfect reference test. J Am Statistical Assoc 1998; 93(443):920-928.
35. Joseph L, Gyorkos T, Coupal L. Bayesian estimation of disease prevalence and the parameters of diagnostic tests in the absence of a gold standard. Am J Epidem 1995; 141(3):263-272.
36. Walter SD, Irwig LM. Estimation of test error rates, disease prevalence and relative risk from misclassified data: a review. J Clin Epidemiol 1988; 41(9):923-937.

CHAPTER 4

Primer on Clinical Trials

Robin S. McLeod

The randomized controlled trial (RCT) is accepted as the best trial design for comparing two medical therapies. It is similar to the experiment performed by the scientist in the research laboratory, in that it has a rigorous design to minimize random error and systematic error (bias) that otherwise might lead to incorrect conclusions or generalizations about the effectiveness of a treatment. However, unlike the laboratory experiment, it is used to address issues of concern in the clinical domain. Thus, the subjects are human beings and the investigator is the clinician researcher. There are several essential components of the randomized controlled trial. First, subjects are randomly allocated to two groups, usually a treatment group (in which the new treatment is being tested) and a control group (in which the standard therapy or a placebo is administered). Thus, the control group is concurrent, and subjects are randomly allocated to the two groups. Second the interventions and follow-up are standardized and performed prospectively. Thus, hopefully, both groups are similar in all respects except for the interventions being studied. Not only does this guard against differences in variables known to be important, but it also ensures that there are no differences due to other factors that have not yet been identified or cannot be measured.

Why should we perform randomized controlled trials? Certainly, randomized controlled trials are a phenomenon of the latter part of the twentieth century. Previously, most clinicians relied on careful observation to make deductions about the efficacy of a treatment. There are many examples of accepted therapies that were introduced to clinical practice without being tested in a randomized controlled trial including appendectomy, antibiotics for treating intra-abdominal infections and colectomy for ulcerative colitis. What these interventions have in common is that the treatment effects are so large compared to the alternative or no treatment, that their effectiveness is obvious. Even if there were selection biases or other differences between the two groups, the magnitude of the difference is such that most of the observed effect must be attributed to the treatment innovation. Thus, a randomized controlled trial is not needed to confirm these observations. More commonly, in modern medicine, technological developments or surgical interventions lead to small increments in survival or perhaps only an improvement in quality of life. It is necessary, therefore, that extraneous factors are controlled to be certain that the observed difference is indeed due to the treatment. While the randomized controlled trial has been of importance in improving patient management, it does have limitations. First, RCTs tend to take a long time to complete because of the time required for

Surgical Arithmetic: Epidemiological, Statistical and Outcome-Based Approach to Surgical Practice, by Lawrence Rosenberg, Lawrence Joseph, and Alan Barkun. ©2000 Landes Bioscience.

planning, accruing and following patients and finally analyzing results. As a consequence, results may not be available for many years. Second, clinical trials are expensive to perform, although their cost may be recuperated if ineffective treatments are abandoned and only effective treatments are implemented.[1] Third, the results may not be generalizable or applicable to all patients with the disease because of strict inclusion and exclusion criteria and inherent differences in patients who volunteer for trials. Fourth, in situations where the disease or outcome is rare or only occurs after a long period of follow-up, RCTs are generally not feasible. Finally, the ethics of randomized controlled trials is controversial and some clinicians may feel uncomfortable with randomizing their patients when they believe one treatment to be superior even if that is based only on anecdotal evidence.[2,3]

Like the laboratory experiment, the randomized controlled trial must be carefully planned, rigorously implemented and carefully analyzed for the results to be valid. While certain aspects of the trial will vary depending on the objectives of the study and the treatments being tested, there are common elements in all randomized controlled trials. In this Chapter, the considerations in designing a RCT will be discussed with reference to some of the special issues in surgical trials. Finally, some of the issues in implementing trials including their administration will be discussed.

1. Determining the Question

Like any research, a clinical trial starts with an idea or a hypothesis. However, before embarking on a trial, this vague idea or question must be operationalized into a specific research question which specifies the sample to be studied, the interventions to be investigated, the comparisons to be made and the outcomes of interest. For instance, one might start with the idea "Does a stapled ileoanal anastomosis (IAA) lead to improved outcome in pelvic pouch surgery?" To transform this into a research question, one must specify "in whom will it benefit?" This could be patients with familial adenomatous polyposis or ulcerative colitis or it could even be more restrictive and include only patients with ulcerative colitis with no evidence of dysplasia or a short history of disease. Is it better than what treatment? The control intervention would likely be handsewn ileoanal anastomosis but what ever it is, the technique of both interventions would have to be specified. And, finally, what is meant by improved outcome? The outcome measure could be surgical complication rates, functional results (stool frequency or continence) or quality of life. If so, how will these be measured-with a patient diary or a generic or disease specific quality of life instrument? So, the hypothetical research question, after considering these issues might be "In patients with ulcerative colitis without evidence of dysplasia, does a stapled IAA result in improved quality of life as measured by the Inflammatory Bowel Disease Questionnaire (IBDQ) compared with a handsewn IAA?" Thus, the process of clearly enunciating the hypothesis into a precise question is of great importance. In formulating the question, the investigator must have a thorough knowledge of the literature, be aware of the biological rationale and current status of knowledge in the area, and apply the appropriate methodological considerations. It may or may not be readily apparent whether a proposed trial is feasible. In this Chapter it is assumed that the study design will be a randomized controlled trial, but with deliberation, it is quite possible that another design might be more appropriate or that a randomized controlled trial is not feasible. For instance, in the example mentioned previously, the risk of cancer in the rectal cuff was ignored in the objective. If this were deemed to be the most relevant outcome, a RCT would not be feasible because of the infrequency of the event. Thus, at this stage, the investigator might decide to do a case control study instead.

The three necessary attributes of the research question are that it has biological rationale, clinical relevance and significance, and is feasible. Data from animal studies, and anecdotal reports or case series should be available to provide supporting evidence to justify a randomized controlled trial. Sometimes, there may be a sequence of RCTs to evaluate a particular intervention. For instance, initially a rather small trial which is highly controlled using what might be called a surrogate outcome measure (e.g., physiologic or x-ray measure) to assess outcome might be performed. Generally, few patients would be needed and a large difference in outcome would be expected. If the study were positive, this might lead to a second, larger trial with a more clinically relevant outcome measure. Subsequently, a very large trial might be indicated to assess the effectiveness of the intervention in normal clinical practice.

From a pragmatic view point, it is important that the trial have the above attributes since the researcher will have to convince a research ethics committee of the ethical and scientific soundness of the trial before it will be approved, funding agencies that it is worthy of support and finally scientific journal editors and other health professionals that the results of the trial are significant so that the work should be published and the treatment adopted. In addition to the study being based on sound biological rationale, it must be feasible in terms of the numbers of patients available, the expertise and equipment available and the resources involved. Thus, if the disease is rare or the particular outcome occurs infrequently or after long follow-up, a trial is probably infeasible even if it is scientifically exciting and worthwhile.

There may be additional secondary questions which should be selected as carefully as the primary question. Secondary questions may be of two types: first, the design of the study will allow other variables to be measured and thus other questions to be answered. Second, subgroup analyses of the primary outcome may be performed. Questions related to answering basic mechanisms of disease may also be answered but clinical trials tend to be pragmatic and often are not the best design for answering fundamental questions related to pathophysiology. However, questions related to cost effectiveness are often required by regulatory bodies and may be performed simultaneously with the trial.

2. Selection of Subjects

The subject selection will vary depending on the question being asked, again emphasizing the need for the researcher to have a clearly enunciated question. Like other aspects of designing a clinical trial, the inclusion criteria can be very broad and unrestricted or they can be narrow and very restrictive, or they may lie somewhere in between. For instance, in designing a trial assessing adjuvant therapy in colorectal cancer, one could choose to include patients with all stages of disease (i.e., Stages I-IV) or restrict entry to only patients with Stage III disease only. The advantage of having broad inclusion criteria, and including all patients with colorectal cancers is that the number of potentially eligible patients is larger and secondly, the results of the trial would be more generalizable. The disadvantages and thus, the advantages of the restricted inclusion criteria, are that if patients are less homogeneous and some are less likely to respond to adjuvant therapy, the overall treatment effect will be diluted. Secondly, there is always a tradeoff in the sample size required: although there are more patients eligible with broadly inclusive entry criteria, there is also more variation in these patients and therefore a larger sample size would be required since variation (standard deviation) is one of the variables used to calculate sample size (see Chapter 2.) How does one decide on who to include in the trial? Several factors may have to be considered depending on the trial. In the situation previously posed, the biology of the tumor would dictate that patients with Stage I and IV tumors should be excluded since they are unlikely to benefit from adjuvant therapy. On the other hand, in a trial evaluating the use of surgical drains, it might not be as obvious which patients to include or exclude.

Generally, the inclusion criteria are stated first to define the study population. These are the patients who compose the population who are likely to benefit from the intervention and also in whom a benefit can be detected or measured. Then, the exclusion criteria will define the subgroup of patients who fit the inclusion criteria but who will be excluded. The most common reasons for excluding patients are:

1. patients who are at increased risk for developing adverse events e.g., due to allergy, past or present history of other disease;
2. patients who are at risk for developing conditions which might preclude the ascertainment of the event of interest (e.g., another disease such as another carcinoma which might shorten their lifespan; another medication which cannot be stopped but might be a cointervention; a disability such as intellectual impairment which would preclude completing forms) and finally,
3. patients who might be noncompliant for a variety of reasons including distance from the treatment center.

Both the inclusion and exclusion criteria should be clearly defined and stated in advance. They should lack ambiguity. Thus, for example, rather than stating "subjects will be excluded if they have other comorbidities", the specific comorbidities should be stated and a defined. Thus, exclusion criteria might be renal failure defined as a serum creatinine greater than 150 mmol/l or a history of thromboembolism based on previous history and confirmation with a Duplex ultrasound. The degree of precision with which each criterion is defined may vary depending on its importance and the degree of acceptance of definitions. Thus, the TNM staging system is well accepted for colon cancers and further description would not be required. However, surgeons' opinions vary with respect to what constitutes a rectal cancer (i.e., below the sacral promontory or below the peritoneal reflection etc.) so it would be important to define it in planning a trial of adjuvant therapy for rectal cancer. On the other hand, in this situation it would be less important to precisely define an exclusion criteria of "no history of bleeding disorders" but very important to do so in a trial studying thromboembolic prophylaxis.

3. Allocating Subjects

As stated previously, the hallmark of the randomized controlled trial is that patients are randomly allocated. Randomization is important because it tends to produce study groups comparable with respect to known as well as unknown risk factors, removes investigator bias in the allocation of subjects and guarantees that statistical calculations will be valid. In simplest terms, with randomization each subject has the same chance of being assigned to either the control or treatment group.[4]

A randomization scheme can be simple in its design or more complicated depending on the size and nature of the trial. One essential feature is that randomization should be performed blindly. In other words, neither the subject nor the investigator should know what the assignment will be before the subject's decision to enter the study. If they do, accrual of patients may be biased depending on the patient's and the investigator's prior beliefs about the therapies. This means that randomization based on the patient's birthdate or hospital ID number is inappropriate since randomization would not be performed blindly.

The simplest method for randomizing subjects is by simple coin toss. A more convenient method, however, and also one free of the biases of the coin tosser, is to use a random number generator available in table form in statistics texts or on most computers. For example, subjects could be assigned to treatment group A if there is an odd number and treatment group B if there is an even number or vice versa. Block randomization is a modification of this method. Again numbers are randomly generated and subjects are randomly allocated but only within a small block of subjects. Thus, for a block of 8 subjects, there would be 4 subjects assigned to each intervention, the 4 positions for treatment A selected randomly from the 8 possible positions, and made different in each block of 8 patients. The advantage of block randomization is that it prevents imbalance in the number of subjects assigned to each group which might occur with simple randomization. At all times, there are relatively equal numbers of subjects in each group. Thus, if one is concerned that treatment effectiveness might vary depending on when the subject enters the trial, block randomization would eliminate this as a potential bias. As well, if the study is terminated early, there would be equal numbers in both groups. The usual block sizes vary between 4 and 8 subjects, often with variable block sizes to eliminate the possibility of an investigator being able to guess to which group the next subject would be randomized.

Prior to randomization, subjects are often stratified according to various prognostic variables. Although one expects that randomization will ensure that both groups will be identical, by chance, it is possible that there may be an imbalance in some variables. If that were a variable of little or no prognostic significance, it would be of little importance. On the other hand, if it were of great importance, it could invalidate the results of the trial. Thus, subjects are often stratified prior to randomization to ensure that the groups will be balanced according to the most important prognostic variables. Thus, for a trial of adjuvant therapy in patients with colorectal cancer, patients might be stratified according to site of the cancer (colon or rectum) and stage (II or III). Subjects would then be stratified to one of 4 strata (i.e., Stage II Colon cancer, Stage III Colon cancer, Stage II rectal cancer or Stage III rectal cancer) and randomized to one of the treatment groups within each stratum. Only a few of the most important variables should be used for stratification because the number of strata increases factorially for each variable. With a large number of strata, there may be few subjects randomized into each cell. As stated previously, the purpose of stratification is to ensure balance within the treatment groups. It is not performed so subgroup analyses can be performed at the conclusion of the trial.

In most studies, subjects are allocated equally to the two treatment groups. However, in some, subjects may be allocated unequally, say, in a 2:1 allocation, with 2 individuals being assigned to the treatment group for each 1 individual who is assigned to the control group. Although there is less statistical power with this method and thus, relatively more subjects are required, it may be employed in situations where most patients have a stronger preference for one treatment.[5] They may be more likely to enter the trial if they have a greater than 50% chance of receiving that treatment. This is especially appealing in conditions where there is no effective standard treatment for the disease so the control group will receive a placebo medication and there is a new medication with some promise being tested in the treatment group.

The mechanics of randomization are as important as the method of randomization. A simple method for randomizing patients is to transcribe the randomization scheme initially generated using random numbers to individual cards which are placed in sealed envelopes and opened in sequence when a patient is to be randomized. Alternately, the algorithm may be maintained by a third party. The latter is preferable since it eliminates the possibility of the sequence of envelopes being altered or the envelopes being opened or tampered with either advertently or inadvertently. Additionally, if the third party checks the clinical information before randomizing the subject, it will ensure subjects meet the eligibility criteria and that subjects are randomized into the correct strata. For multicenter trials, often investigators phone a central randomization center to obtain the randomization number. For single center trials or small multicenter trials, the randomization scheme may be maintained by the pharmacy or another independent source. Throughout the study, an independent assessor should monitor the randomization procedures to ensure it is being performed as a priori specified.

4. Describing the Maneuver

The maneuver is the "what", "by whom" and "when" part of the protocol. In this section of the RCT protocol, not only should the treatments or interventions be described but also what baseline assessments and follow-up maneuvers will be undertaken. As well, the investigator should consider potential biases or "things that could go wrong" which might distort the results and devise strategies to eliminate or minimize these risks. Usually the investigator has already decided on the interventions before beginning to plan the trial because they led to his/her desire to "do" a trial. However, the maneuver must be described in detail so that it can be replicated precisely by a clinician wishing to implement the results of the trial and, of more urgent concern, by other investigators participating in the trial.

4.a. The Interventions

There must be careful consideration of the choice of both the control and the treatment interventions. For medical trials, a placebo control group is indicated if there is no standard therapy whose effectiveness has been clearly established, preferably in randomized controlled trials. If there is a standard therapy, then this would be administered to the control group. The experimental therapy should be chosen based on current understanding of human structure, function, pharmacology and clinical practice. In most instances, there will be evidence from basic or animal research plus case series in humans that suggest the intervention may be of benefit. In addition, the intervention should be accessible, affordable and acceptable to patients.

Standardization of the intervention (operation) is of particular concern in surgical trials. In pharmaceutical trials, the dosage can be standardized and compliance can be measured, so that the prescribing physician is not a variable. In surgical trials, however, standardization of a procedure may be difficult because surgeons vary in their experience with and their ability to perform a surgical technique, there may be individual preferences in performing the procedure, and there may be technical modifications as the procedure evolves.

Although a certain minimum amount of standardization is mandatory in any surgical trial, the amount may vary depending on the question being asked in a similar way that compliance may vary in a pharmaceutical trial. Thus, a trial where surgical technique is highly controlled (i.e., few surgeons who are experts in the field using similar technique) is analogous to the medical trial where only compliant patients are randomized. This may be viewed as an efficacy trial. Where the surgical maneuver is less controlled, it may be viewed as an effectiveness trial. In other words, the question is: "Is this surgical procedure effective when performed by many different surgeons without special expertise with this procedure?" Depending on the design of the trial, the generalizability of the results will vary. However, no matter what strategy is adopted, it is essential that the investigator describe how the procedure was performed and by whom so the reader can determine whether the results are applicable to his/her practice and that the technique can be replicated.

Standardization of the control operation is as important as it is for the new operation. It is often easier to do so since the procedure may be more "mature" and the technique may be more standardized. Other aspects of care which are deemed important such as prophylactic antibiotic usage, postoperative care etc. should also be standardized. There are several strategies that may be instituted to ensure that a certain uniformity in surgical technique is achieved and that the effect of the surgeon as a variable is minimized. First, participation may be limited to a small number of surgeons or only those where there is documentation that a satisfactory number of procedures have been performed or that they can perform it adequately. This will lead to increased standardization of the technique although the results may be less generalizable. Second, investigators can meet prior to the trial and reach a consensus on the technique for the critical aspects of the procedure. Third, teaching sessions may be held and manuals produced which describe and illustrate the accepted technique. Finally, patients should be stratified by surgeon or center as discussed previously. There may still be variation in how the procedures are being performed but it will ensure that there is not an imbalance between groups.

During the trial, compliance in taking medication must be monitored at regular intervals, to ensure that compliance is adequate and to allow the reader to assess the validity of the results. For instance, if the trial results were "negative", (e.g., a result centered near a zero difference in success rates and with a narrow confidence interval) but mean compliance was only 50%, the reader would be less convinced of the ineffectiveness of the treatment than if compliance were closer to 100%. Typically, for medical trials, compliance is measured by having patients return bottles of unused pills or possibly, by measuring serum levels of the medication. Both have their limitations. For surgical trials, on going measurement of compliance is unnecessary but compliance in performing the procedure adequately (quality assurance) can be audited by reviewing videotapes of procedures (endoscopic, laparoscopic), pathological examination of resection margins, nodes etc. or x-ray evidence of vessel patency. Where results are suboptimal, performance may be improved by providing feedback to surgeons.

4.b. Potential Biases

4.b.1. Lack of Blinding

Bias may occur due to lack of blinding of patients and/or investigators. The placebo effect is well documented and is of particular concern in surgical trials where it is known that surgery often has a significant placebo effect.[6] However, the importance of blinding may vary depending on the trial and the primary outcome measure. For instance, if the outcome measure is all cause mortality, then even if there is no blinding, the results will probably not be biased provided that there is a uniform search procedure for deaths across all treatment arms. On the other hand, if the primary outcome is quality of life, then lack of blinding may potentially bias the results. Investigators should therefore make every effort to minimize this bias, if blinding of the treatments cannot be ensured.

In trials comparing medical and surgical therapies, blinding may be impossible since sham operations are generally unethical. Even in trials comparing two surgical operations, there may be difficulty blinding patients and investigators if the incisions differ or the operations differ in magnitude. Creative ways may be required to minimize blinding such as applying large, uniform dressings to the wounds.[7] To minimize the lack of blinding, a "hard" outcome measure such as recurrence of disease or death may be chosen. However, more often the important outcome is a change in symptoms or quality of life, and outcome may be biased by knowing which treatment group the subject is in. In these situations, the potential for bias may be minimized if a hard outcome measure is also assessed and it is found to correlate with the patient's subjective assessment.[8] If possible, assessments may be performed by an independent assessor who is unaware of the treatment group that the patient is in and the outcome measures have been explicitly defined a priori. A blinded panel may also be used to review results of tests using these criteria but without knowledge of the treatment allocation.

4.b.2. Contamination

Contamination refers to the inadvertent administration of the experimental therapy to the control subjects, and vice versa. Most commonly, in medical trials, it occurs when the control subjects receive the treatment medication. In surgical trials, it may occur due to subjects crossing over to the other group and receiving the experimental treatment. Its effect is that it may spuriously reduce the outcome differences between the experimental and control subjects. Close monitoring is necessary to ensure that it does not occur either advertently or inadvertently.

4.b.3. Cointervention

Cointervention refers to the additional diagnostic or therapeutic maneuvers that are carried out on an experimental or control patient. For instance, subjects in trials evaluating NSAIDS should be asked to refrain from taking ASA even if it is taken for another indication. To minimize bias due to cointervention, there should be standardization of the interventions plus the ancillary care. In surgical trials, for example, measures that are part of the peri-operative care that may impact on outcome, such as antibiotics, and ICU care should be standardized. As well, attending physicians and the subject should be alerted to medications or treatments that should not be taken unless absolutely indicated. If, however, these treatments are taken, it should be documented.

4.c. Baseline and Follow-up Maneuvers

Similar baseline and follow-up procedures must be performed on both groups of subjects. They should also be performed at the same intervals and without knowledge of the subject's treatment allocation. Baseline procedures serve 3 purposes. First of all, some tests may be necessary to determine whether subjects fit the inclusion and exclusion criteria. Secondly, baseline values of the outcome measures are necessary to ensure comparability of the two groups at baseline in order to assess differences at the start and the completion of the trial. Thirdly, some tests may be used throughout the trial to monitor toxicity.

Subjects should be followed at specified intervals throughout the trial with the frequency of follow-up visits depending on the duration of the trial and the anticipated timing of outcome and adverse events. The intervals between follow-up may vary. For instance, if one anticipates that adverse events such as allergic reactions to medication or postoperative complications are more likely to occur early in the trial, visits may be more frequent then than later in the trial. Follow-up visits serve three purposes: to assess outcome, toxicity and side effects and compliance. Appropriate tests, procedures and questionnaires should be completed at each visit to assess each of these. In some trials, the outcome event may occur between scheduled visits (e.g., symptoms suggestive of recurrent cancer) and if so, follow-up procedures to document the outcome event will have to be performed then. Similarly, follow-up procedures should be specified in the protocol in case subjects develop adverse effects between follow-up visits.

5. Measuring Outcome

5.a. Assessing Treatment Effectiveness

Assessing outcome is an important part of any clinical trial. Assuming that the subjects are similar in every way other than the treatment they receive, then a difference in outcome is used to make inferences about the effectiveness of the treatment. There are a variety of outcomes that can be measured. Traditionally, in surgical trials, outcome has been assessed in terms of mortality or survival and complication rates. Thus, operative mortality or long term survivorship have been of primary concern. However, death is a relatively infrequent occurrence following modern surgery. Secondly, in most instances, the primary indication for surgery is not prolongation of life. For instance, hip or knee replacements, vascular procedures for claudication, prostate surgery and surgery for reflux esophagitis and inflammatory bowel disease generally do not prolong the patient's life but do improve his/her quality of life. Similarly, while the complication rates following surgery are of concern, they may be less important than other patient centered outcomes especially in the long term. As a result, there is now an emphasis on assessing functional status and health related quality of life. While functional status tends to be more relevant to patients than the other traditional measures, it may not necessarily correlate with quality of life. For instance, the quality of life of patients who have had a pelvic pouch does not necessarily correlate with their stool frequency despite the latter being the primary outcome measure used to evaluate outcome after this operation. Similarly, mobility and pain are not the sole determinants of quality of life in patients who have undergone a joint replacement. Health related quality of life, while understood by most people, is hard to define. The World Health Organization has defined health as "a state of complete physical, emotional and social well being and not merely the absence of disease".[9] This definition of health related quality of life has been adopted by many.

Although several outcomes may be assessed, generally one outcome is chosen as the primary outcome measure and all others are secondary. The primary outcome is chosen because it is deemed to be the most clinically relevant. The sample size is based on its rate of occurrence and conclusions regarding the effectiveness of the treatment are based in large part on this variable. Outcome may be measured in three different ways: first, dichotomously or as an event in which something either occurs or does not occur. Examples would be perioperative mortality, and complication rates. Secondly, it might be measured as a continuous response variable which would be the case in measuring quality of life or assessing change in a physiological variable such as blood pressure. Third, outcome might be measured as the time to an event, such as failure, death or recurrence. Examples would be time to recurrence of cancer or Crohn's disease.

Of chief concern in selecting an appropriate outcome is that it should be clinically relevant and responsive to treatment. Thus, outcomes which are of importance to patients should be chosen. For instance, in a trial comparing two treatments for gastroesophageal reflux, the relief of heartburn and other symptoms would be more relevant to patients than the endoscopic appearance of the esophagus. The latter is sometimes referred to as a surrogate measure. Generally, one should try to avoid using a surrogate measure (such as a laboratory or x-ray test) as the primary outcome measure although these may be used as secondary outcome measures in support of the primary outcome. In addition to being clinically relevant, the outcome should be responsive to change.

It is because of the belief that important outcomes are those that are important to patients that quality of life is often chosen as the primary outcome measure. In addition to being responsive or sensitive to change, outcome measures should be reliable and valid.[10] Reliability refers to reproducibility. Thus, if the instrument is administered in the same circumstance to a patient whose status has not changed, the same results should be obtained on each occasion. Validity refers to whether the instrument is measuring what it is intended to measure. With quality of life assess ment, there is usually no "gold standard" against which the instrument can be assessed. Thus, construct validity is established by showing that the instrument gives expected results in comparison to other measures when administered to the same patient.

There are a variety of instruments available to measure quality of life. They can, in general, be classified as psychometrically and utility based measures.[11] Psychometrically based measures attempt to quantify quality of life using a range of questions from the various domains being assessed. The ratings or scores of the individual items are usually summated to give an overall measure of quality of life. There are both generic as well as disease specific psychometric instruments. Generic measures have been designed to be applicable to individuals with a broad range of diseases and impairments, undergoing varied treatments. Their usefulness is that they are applicable to a wide range of groups and therefore quality of life between these groups can be compared. The disadvantage is that they may lack the sensitivity to detect small but clinically important differences in a particular group of patients. Other advantages of the generic instruments are that they often have been used extensively and therefore their validity and reliability have been well established in varied populations, they can detect and measure unexpected treatment affects and they can be useful in cost-effectiveness studies and health policy analysis. Some examples of generic instruments are the Medical Outcomes Trust SF-36,[12] the Sickness Impact Profile,[13] and the Nottingham Health Profile.[14]

Disease-specific instruments have been designed to measure those areas of quality of life of importance to specific patient populations, and so may be more responsive to small, clinically important changes and may better discriminate between individuals within the population. They may also appear more relevant to clinicians and patients. However, they tend to have less established reliability and validity than generic measures and cannot be used to compare different populations to whom disease-specific instruments are not applicable. There are a wide range of instruments available in most disciplines.[15]

The alternative to measuring quality of life psychometrically is using utility based measures.[16-18] Utilities represent individual's preferences for a given state relative to death or perfect health. Complete wellness is given a utility value of 1.0 and death 0. A health state less than completely well is given a value between 0 and 1.0. Utilities may be assigned using either a decomposed or holistic approach. In the decomposed method, an individual is asked to rate his/her functioning in a number of health domains or attributes. For each specific category score a utility value has been previously generated from a defined population and they can then be combined to obtain a total score for that individual. In the holistic approach, individuals assign utilities to various health outcomes taking into consideration all aspects of quality of life. Utilities may be generated using either the standard gamble method or time tradeoff technique and may vary from 0 to 1.0.[16] In the standard gamble, a utility is calculated based on how much risk the patient would be willing to take to have normal health rather than his/her present health state, while in the time tradeoff, the utility is calculated based on how many years of life the patient would be willing to give up. Utility assessments tend to be less sensitive in detecting differences or changes than psychometrically based measures. As well, if the holistic approach is used to generate utilities, it is not possible to discern which domain or aspect of health is affected. Their main use has been in the field of health economics and policy making in performing cost utility studies.

5.b. Assessing Side Effects and Toxicity

In addition to assessing treatment effectiveness, toxicity or adverse events related to treatment must be assessed. Like other aspects of the trial, decisions should be made a priori (i.e., before the trial) as to what adverse events will be assessed and how they will be measured. If the trial is assessing a surgical intervention, it is obvious that perioperative mortality and complications will be measured. There are a variety of ways of reporting them. The investigator may simply decide to record all complications and report them as a single figure. However, since complications tend to vary in severity, it may be more useful to record them individually or group them as major or minor complications. One might also assume that some complications (e.g., postoperative pneumonia, DVTs, urinary tract infections) will be the same in both groups irrespective of the treatment and only those related to the intervention (e.g., intestinal anastomotic leak rate or joint dislocation rate) will be recorded and reported. Whatever is decided upon, those complications that are clinically important because of their severity or frequency should be included and decisions related to how they will be measured should be made a priori so they can be assessed objectively and similarly throughout the trial.

As stated previously, some laboratory tests or x-rays may be performed routinely throughout the trial to monitor for side effects or toxicity which would be predicted to occur in a proportion of patients. Thus, for example, patients receiving Imuran would require frequent monitoring of their white cell count.

6. Analyzing the Data

A statistician should always be a part of the investigative team involved in a clinical trial and play a major role in the design, monitoring and analysis of the trial. Thus, only the general concepts of data analysis will be discussed in this section. (See Chapter 2 for more details.)

As a general rule, all patients who are entered into the trial should be included in the analysis. If not, there is a risk that the study groups may not be similar. Secondly, random allocation is the basis upon which statistical inference is made.[4] Finally, from a pragmatic point of view, most journal editors will not accept an article unless such an intention to treat analysis has been performed. An effectiveness or intention to treat analysis includes all subjects who are randomized and analyzes the patients in groups to which they were randomized, even if they in fact received the other intervention or were noncompliant. The idea is to mimic as closely as possible what might happen in clinical practice. An efficacy analysis includes only subjects who fit the entry criteria, receive the treatment to which they were allocated and were compliant with it. Despite precautions to ensure there are no protocol deviations, some almost always occur. Thus, it is customary to perform both analyses efficacy and effectiveness but the effectiveness analysis is usually the primary one upon which conclusions are made. If an investigator wishes to determine whether a treatment is effective in optimal conditions (i.e., perform an efficacy trial), this should be determined a priori and the design and performance of the trial may differ from that of an effectiveness trial. It is not done by excluding noncompliant subjects at the end of the trial.

Problems with data analysis are minimized if there is rigorous monitoring of the trial to minimize protocol deviations. Some of the most common protocol deviations are including subjects who do not meet the entry criteria, subjects not receiving the treatment to which they were randomized, poor subject or investigator compliance, subjects "dropping out" and failure to make a final outcome assessment. Ensuring that subjects meet the entry criteria can be accomplished by making certain that all of the information is available to determine eligibility prior to randomization of subjects. Sometimes errors do occur if patients are randomized in emergency situations or if patients are randomized before laboratory, pathology or radiology results are reviewed by a study investigator or adjudication committee. These errors should be minimized to the extent possible. However, excluding patients who do not fit the entry criteria from the analysis is more acceptable than exclusions for other reasons. The other causes of protocol deviations should be minimized by rigorous follow-up of subjects throughout the trial to ensure they receive the correct treatment, they are compliant with it and have a final outcome assessment. Generally, these subjects should be included in the analysis in the group to which they were randomized. Peto has argued that if the failure to complete the trial or receive the correct treatment occurred due to chance, there should be an equal proportion of patients with protocol deviations in each group.[15] If so, by including all patients in the analysis, the treatment effect, if there is one, may be reduced, but otherwise the results will not be biased. On the other hand, if the failure is due to an effect of the treatment (e.g., toxicity or intolerance of the treatment), omission of these patients from the analysis might bias the results in favor of one of the treatments.

Subjects may not have a final outcome assessment for a variety of reasons including:

1. they developed a side effect and had to be withdrawn,
2. they refused further follow-up or
3. the final assessment was not performed adequately.

Prior to analyzing the data, it is wise to develop "rules" as to how these situations will be handled. For instance, if the subject developed a side effect necessitating withdrawal of therapy, this may be considered a treatment failure. If the outcome assessment is based on a continuous variable (such as an activity index, or HRQL score), then the last outcome assessment performed before the subject was withdrawn because of the side effect may be used. Hopefully, the number of subjects who "drop out" because they refuse follow-up can be minimized. Even if subjects refuse treatment or stop treatment during the trial, they may be counted if they will agree to having the final outcome assessment performed or this information is available. For example, for a trial where survival is the main outcome, this information may be available even in the noncompliant patient. Finally, there may be some situations where the final outcome assessment is not available. Examples are situations where x-rays or procedures are incomplete or inadequate (venograms for the assessment of DVTs and colonoscopies assessing recurrent Crohn's disease) or the final outcome assessment was not as described in the protocol (subjects who had surgery for recurrent Crohn's disease without having the required investigations prior to surgery). These might be situations where an independent adjudication committee can rule whether the subject has or has not had an event, or where results from other tests (e.g., Duplex ultrasound for DVT documentation) may be deemed acceptable.

Again, these cases should be reviewed or "rules" should be made prior to analysis of the data and certainly before there is unblinding of investigators performing the analysis.

In clinical trials, often the statistical analysis is relatively straightforward because there is an assumption that the two groups are similar at baseline due to the trial design. Thus, more complicated statistical methods (such as multiple regression analyses) may not have to be performed to correct for possible imbalances in the groups. However, it is important to demonstrate that indeed the groups are similar and this is usually done by constructing a table showing the characteristics of the two groups including the important baseline prognostic variables (for instance, in a trial comparing cancer treatments, the proportion of patients with each stage of disease would be shown). If there is any doubt about the balance in the groups or if there are any variables other than the treatments that may affect the outcome (for example, age) this may investigated using regression techniques, (see Chapter 5). However, should this occur with an important prognostic variable, it certainly will decrease the confidence in the conclusions of the trial. Depending on the outcome variable, different statistical methods will be used to draw inferences about the differences between the treatments. (See Chapters 2, 5, and 6.)

Secondary and subgroup analyses are often performed. Subgroup analyses consist of dividing patients into different subgroups and comparing treatments within each subgroup. While these analyses are interesting, generally they should be restricted to priori identified subgroups where treatment differences could vary from the overall average effect. The danger of indiscriminately comparing all possible subgroups is that spurious relationships may be "found" due to chance alone. To minimize this risk, Investigators should decide, prior to unblinding, which subgroup analyses make biological and clinical sense and perform only these analyses. This is preferable to simply analyzing the data in as many ways as possible in the hope of finding a statistical difference in one of the subgroups (so called "data dredging").

Finally, many clinical trials will be designed so that an interim analysis is performed. An interim analysis is usually performed in case an early dramatic difference in the two treatment groups is observed suggesting that the trial can be stopped early. The advantages of performing an interim analysis, with the possibility of early termination of the trial, is that resources are conserved and theoretically, other patients with the same condition will be able to benefit from the superior treatment as soon as possible. The latter has been a major issue related to trials involving AIDS patients.[19] The potential disadvantages of stopping a trial early include the possibility that an early difference in the treatments may have occurred by chance, and the sometimes difficult task of convincing the clinical community of the validity of the results in an "incompleted" trial. While statistics plays an important role in interim analysis decisions, it is of utmost importance to ensure the safety of all patients. For example, one may start an interim analysis by comparing the two treatment groups in terms of the primary outcome, usually by calculating a confidence interval and seeing if clinically important difference in either direction has been strongly demonstrated. Next, one would compare the rates of all adverse events, to assess safety. Finally, if it appears that the data suggest an advantage for one group, the plausibility should be checked against background knowledge. For example, if the placebo group appears superior to a new technique with strong supporting background evidence, one may decide to continue the trial until stronger evidence emerges. Since interim decisions involve a multitude of considerations, they are often difficult.

Like all other aspects of the clinical trial, decisions regarding an interim analysis should be made early and while all investigators are blinded with respect to the data. These decisions include how many interim analyses will be performed, when they will be performed and what actions will be taken with the results obtained. Then, the analysis should be performed by the statistician and, unless the monitoring committee decides the trial should be terminated, all other investigators should remain blinded with respect to the data and the results of the interim analysis.

4

An interim analysis may not be desirable or feasible in all trials. For instance, in trials where follow-up is long relative to the accrual period, results may not be available for analysis until all of the patients have been accrued. Most commonly, only one interim analysis is performed and it is done when approximately half of the subjects have been accrued. While interim analyses are usually performed to determine whether the trial can be stopped early, another reason to do one is if there is great uncertainty about the event rates used to calculate the sample size initially. An interim analysis may be performed with the idea that the sample size for the trial may be adjusted depending on the observed event rates.

7. Estimating the Sample Size

There are numerous articles and statistical textbooks which outline specific formula and methods for estimating sample size and computer programs which will perform the calculations.[20] Thus, a detailed explanation of the various formulae for estimating sample size is beyond the scope of this paper. In addition, for most trials, a statistician should be consulted for assistance.

The sample size estimate is based on the primary outcome measure. Thus, before calculating sample size, one must decide what the primary outcome measure is. Sample size may be checked or adjusted to ensure that it is large enough to answer the secondary questions but since they are secondary objectives, it is less important that there is enough precision to definitely answer these questions.

Generally, there are 5 variables which affect the sample size: the event rates or mean outcomes in the groups, the standard deviation or variance of the primary outcome in the groups, the difference in the outcomes between the two interventions, the desired precision in estimating this difference, and the degree of confidence in the results (usually 95% confidence intervals are used.) Usually, the expected outcomes and variance can be estimated using data from published reports (even case series) or pilot studies (See Chapter 2 for details). Determining what constitutes a clinically important difference between the two treatments is usually a clinical decision made by the clinician investigator.[21] The larger the expected difference in event rates between the two groups, the less precision is required in order to find the difference so the smaller the sample size required.

For trials where a survival analysis will be performed, the length of time required to accrue and follow patients must also be considered in the sample size estimate.

8. Ethical Considerations

Some have argued that randomized controlled trials are never ethical.[2,3] They would argue that generally, one would not do a randomized controlled trial unless there was some evidence to suggest that one treatment might be better than another. If so, then it is unethical for a physician to withhold that treatment from his/her patient since the good of the individual patient should be his/her primary concern. However, others recognizing the need for society to determine whether treatments really are effective and that this can be done best with randomized controlled trials, state that randomized controlled trials are ethical if there is genuine community equipoise even if there is not individual equipoise.[22] Freeman has argued that if less than 70% of medical experts consider one treatment option to be superior than the other, than a trial is ethical and most reasonable patients would accept randomization.[22]

Most institutions have ethics review committees which have stringent policies for human experimentation and investigators need to be aware of these policies. Generally, for a trial to be ethical, subjects must not be exposed to undue harm or risk and they should be fully informed of both the benefits and harms of the trial. They should give consent without any pressure and all reasonable efforts should be made by the investigator to maintain confidentiality. In addition, however, the trial itself must be clinically relevant. If not, it is not ethical to solicit patients' participation even if there is little risk involved. Investigators must not have any conflict of interest in performing trials. Trials assessing new technologies are often industry driven and investigators may be financially rewarded for each patient that is accrued into the study. In such situations, the surgeon may not be objective in discussing the trial with the patient and obtaining his/her consent. Therefore, it may be preferable to have a third party involved in this aspect of the trial. Some universities have developed guidelines for obtaining consent in these circumstances. Finally, even before embarking on such a trial, investigators should feel that the trial has merit and is answering an important clinical question and that there is uncertainty with respect to the effectiveness of the treatment options. Financial gain should not be the reason for participating in a trial.

It is often said that it is not ethical to randomize a patient in a trial where one treatment option is surgical. In fact, what is more accurate is that it may not be feasible as will be discussed subsequently. As long as there is community equipoise, the patient is well informed and an informed consent is obtained, a trial comparing a surgical to a medical alternative is quite ethical. Successful completion of many important trials comparing medical and surgical therapies is witness to this.[23-25]

It is more likely that in surgical trials the subject may not be able to sign an informed consent. Examples are trials performed in emergency situations, on trauma victims or patients in the ICU. It may be necessary in these situations to rely on third party consent. This issue is complex and local ethics committees may vary in their opinions regarding third party consent.

9. Administrative Issues

9.1. Feasibility of Trials

In order to successfully complete a clinical trial, there must be adequate eligible patients and facilities; committed and experienced investigators and adequate financial resources. Although these are issues related to the implementation of the trial, they must be addressed in the planning or design phase of the trial.

9.1.a. Patient Recruitment

Patient recruitment is often difficult and frequently overestimated by investigators. If recruitment is a problem, successful completion is more likely if a multicenter rather than a single center trial is undertaken. Adequate funding to allow hiring of research assistants who can ensure that all eligible patients are approached to participate is mandatory. In the planning stages, a realistic estimate of patient accrual should be made taking into consideration the number of patients who have the condition that are seen at each institution each year, the proportion who would be eligible and the proportion who would consent to participate. Patient participation can be quite variable, being as low as 10%, depending on the treatments, the intensity and length of the follow-up and the invasiveness of follow- up investigations. The number of participating centers and the duration of patient accrual may be determined based on these estimates.

Patient recruitment to surgical trials may be even more difficult than to medical trials. Patients often have strong preferences for one therapy or another and refuse to participate in a trial. As well, in a medical trial, patients may be offered randomization to one arm or the other with the possibility that at the conclusion of the trial they can receive the more efficacious treatment if the disease is not progressive and the treatment is reversible. Surgical procedures, however, are almost always permanent and therefore patients will not have an opportunity to benefit from the results of the trial. Accrual may also be hindered because there is a lot of emotion associated with surgery itself and patients often are preoccupied with these other issues. As well, they may be reluctant to leave the decision as to whether they will have surgical therapy to chance alone. Other issues such as patients feeling like a "guinea pig" are common to both medical and surgical trials, but again may be of greater significance in surgical trials.

With medical trials, because of restrictions imposed by regulatory agencies, it is unusual that the new therapy is available outside of the trial. This may be an incentive for patients to participate in the trial. Since there are usually no restrictions on surgical therapies, there may be less incentive for patients to participate in a surgical trial. However, the investigators may consider not performing the new surgical procedure outside the trial. This decision has to be made considering the resources available as well as the ethics of restricting the procedure to the trial.

9.1.b. Administration

The team of investigators and the administrative structure of the trial may vary in composition and size depending on the size of the trial and the number of centers involved. For a single center trial, there will be at least one investigator and a research assistant. The investigator will oversee all aspects of the trial while the research assistant will be involved in the day to day coordination of the trial. A well qualified research assistant is essential. In addition, a data manager who can create and maintain a database and oversee data entry is essential. Lastly, a statistician who can perform the sample size calculation, develop the randomization scheme and ensure randomization is performed correctly and analyze the data should be part of the team of investigators. It is important that all are involved in the planning of the trial, especially the statistician.

For multicenter trials, an investigator and research assistant at each site should be part of the investigative team. There may be various committees including a steering committee made up of a smaller number of individuals who will play a more major role in decision making and ensure that the trial proceeds as planned. As discussed in section 6, most trials include an independent safety committee whose duty is to monitor adverse events.

9.1.c. Data Management

The importance of accurate data collection cannot be too highly stressed. The role of the data manager will generally be to oversee all aspects of data collection and entry. He/she will develop a database which will be used to enter the data and subsequently used for data analysis. He/she will supervise data entry personnel. There are many different personal computer data bases available, each with advantages and disadvantages and the choice is ultimately up to the investigators and data manager. However, this aspect of data management has been simplified greatly in recent years with the ready availability of powerful data bases.

Since the results of the trial are dependent on the accuracy of the data, multiple strategies should be instituted to minimize the risk of data entry errors. Some trials now have the research nurses enter the data directly into the computer to eliminate the risk of transcription errors. Others use scanning devices to transfer data from the hardcopy into the data base. There are obvious advantages to these but cost may be an issue for smaller trials. In most trials, however, data will be collected on data forms and then transferred into a data base. Generally, it is recommended that data should be double entered, preferably by two different data entry people or at least on two separate occasions. For the most important variables (such as the outcome), additional maneuvers to ensure accurate data entry may be performed. These may include an audit by the statistician or principal investigator comparing the data entered into these fields with the data entered on to the data form.

Conclusions

As the title of this Chapter implies, this is simply an overview of some of the issues related to the design and execution of a randomized controlled trial. While this book is generally aimed at the surgical investigator performing surgical trials, the principles in designing a clinical trial are essentially the same irrespective of the treatments being evaluated. The point to be emphasized is that careful planning and then rigorous monitoring of the trial are essential in order to ensure that results are unbiased and conclusions can be made about the relative effectiveness of the treatments. A dedicated team of investigators with expertise in the clinical and methodological domains, adequate funding and eligible patients are some of the more important requisites needed to successfully complete a trial. However, the time, effort and resources are usually worth it because of the potential impact on clinical practice.

Selected Readings

1. Detsky AS. Are clinical trials a cost effective investment? JAMA 1963; 262:1795-1800.
2. Hill A. Medical ethics and controlled trials. Br Med J 1963; April:1043-1048.
3. Hillman S, Hellman DS. Of mice, but not men. Problems of the randomized controlled trial. N Engl J Med 1991; 324:1585-1589.
4. The Randomization Process In: Clinical Trials, 2nd Edition, 1985 Editor: Friedman, Burburg CI, Dimeto DL.
5. Peto R, Pike C, Armitage P et al. Design and analysis of randomized controlled trials requiring prolonged observation of each patient. I. Introduction and Design. Br J Cancer 1976; 34:585-612.
6. Dimond EG, Kittle CF, Crockett JE. Evaluation of internal uammary artery ligation and sham procedure in angina pectoris. Circulation 1958; 18:712-713.
7. Majeed AW, Troy G, Nichol JP et al. Randomized, prospective, single blind comparison of laparoscopic versus small incision cholecystectomy. Lancet 1996; 347:989-994.
8. Spechler SJ et al. Comparison of medical and surgical therapy for complicated gastroesophageal reflux disease in veterans. N Engl J Med 1992; 326:786-792.
9. World Health Organization: The First 10 Years of the World Health Organization. Geneva, WHO 1958.
10. Guyatt GH, Feeney DH, Patrick DL. Measuring health related quality of life. Ann Int Med 1993; 118:622-629.
11. Patrick DL, Deyo TA. Generic and disease-specific measures in assessing health status and quality of life. Med Care 1989; 27:S217.
12. Ware JE, Sherbourne CD. The MOS 36-Item Short-Form Health Status Survey (SF-36). 1. Conceptual framework and item selection. Med Care 1981; 19:473.
13. Bergner M, Bobbitt RA, Carter WB et al. The sickness impact profile: Development and final revision of a health status measure. Med Care 1981; 19:787.
14. Jenkinson C, Fitzpatrick R, Argyle M. The Nottingham Health Profile: an analysis of its sensitivity in differentiating illness groups. Soc Sci Med 1988; 27:1411.
15. Quality of life and pharmacoeconomics in clinical trials. 2nd edition, 1996. Editor: Spuker B. Publisher: Lippincott-Raven, Philadelphia.
16. Torrance GW. Measurement of health state utilities for economic appraisal: A review. J Health Econ 1986; 5:1.
17. Torrance GW, Thomas WH, Sackett DLS. A utility maximization model for evaluation of health care progress. Health Res 1972;7:118.
18. Raiffa H. Decision analysis: introductory lectures on choices under uncertainty. Publisher: Addison Wesley. Reading Massachusetts, 1996
19. Merigan T. You can teach old dogs new tricks. How AIDS trials are pioneering new strategies. N Engl J Med 1990; 323:1341-1343.
20. Lachin J. Introduction to sample size : Determination and power analysis for clinical trials. Con Clin Trials 1981; 2:93-113.
21. Naylor CD, Llewellyn-Thomas HA. Can there be a more patient-centred approach to determining clinically important effect sizes for randomized treatment trials? J Clin Epidemiolo 1994; 47:787-795.
22. Freeman B. Equipoise and the ethics of clinical research. N Engl J Med 1987; 317:141-145.

23. North American Symptomatic Caroted Endartertectomy Trial Collaborators. Beneficial effect of carotid endarterectomy in symptomatic patients with high grade carotid stenosis. N Engl J Med 1991; 325:445-453.
24. The EC/IC Bypass Study Group. Failure of extracranial intracranial arterial bypass to reduce the risk of ischemic stroke: results of an international randomized trial. N Engl J Med 1985; 313:191-200.
25. Fisher B, Bauer M, Margolese R et al. Five year results of a randomized controlled trial comparing total mastectomy and lumpectomy with or without irradiation in the treatment of breast cancer. N Engl J Med 1985; 312:665-673.

Linear and Logistic Regression Analysis

R. Platt

1. Introduction

Univariate statistical techniques, which describe or draw inferences about the characteristics of a single variable or measurement, are limited since we are often interested in drawing inferences relating to two or more variables. For example, consider the data given in Table 5.1. One hundred twenty three patients requiring hernia operations were assigned to either conventional hernia repair surgery or laparoscopic repair surgery. The researchers were interested in determining whether the new surgical method reduced the total number of convalescence days. Other variables that could influence the number of convalescence days were the size of the hernia (large or small), the sex of the patient, the baseline health score, the number of years of smoking, occupation (active or sedentary) and whether the patient received government compensation for the surgery (CSST in the table). Eleven patients had incomplete information for some of these variables, so our analysis is restricted to the 112 patients with complete data. Thus, we assume that the missing data are not different in any important respect to the nonmissing data, so omitting cases with missing data causes no bias. While this assumption is necessary for our simple techniques, multiple imputation techniques and other methods have been developed to handle missing data if the assumption is not reasonable.[1] Throughout this Chapter, we will refer to this example and demonstrate how the data can be analyzed using the techniques described here to understand the variability in the number of convalescence days.

The simple t-test and the confidence interval for a difference in means (described in Chapter 2) compare the measurements for an outcome of interest in two different groups. Often, however, researchers are interested in comparing results between several groups, or determining the associations between an outcome and other continuous variables. The correlation coefficient (see Chapter 2) measures the linear relationship between two variables, ranging from 1 (a perfect positive linear relationship) through zero (no linear relationship) to -1 (a perfect negative linear relationship). However, while it provides the strength of the association, it cannot be used for prediction and is restricted to linear relationships between two variables.

Regression models generalize both the correlation coefficient and the t-test. Simple linear regression models can be used for inferring the relationship between two variables, and prediction of the outcome variable based on the other is possible. Simple regression can be further generalized to multiple regression, where we simultaneously

Surgical Arithmetic: Epidemiological, Statistical and Outcome-Based Approach to Surgical Practice, by Lawrence Rosenberg, Lawrence Joseph, and Alan Barkun. ©2000 Landes Bioscience.

consider more than one possible predictor variable, and logistic regression, where we allow response variables that are dichotomous, rather than continuous. Analysis of variance models, or ANOVA, are another related approach to generalizing the t-test to multiple groups, although in fact one can devise multiple regression models that are equivalent to carrying out an ANOVA.

A common thread runs through all of the procedures discussed in this Chapter. The primary interest is to explain the variability in the response by taking into account the structure of its relationship with the possible predictors. If division into groups or the use of an independent continuous regression variable can explain a substantial part of the total variability in the response variable, we can develop a much better understanding of the response variable and more precise predictions of the responses for future individuals.

This Chapter will first describe simple linear regression and how it is used, along with some of the diagnostic checks to evaluate the fit of the model to the data. Multiple regression, the generalization of simple linear regression when more than one independent variable is used to predict the outcome variable, as well as model selection techniques will follow in Section 3. Model selection is important for deciding which model or models to use. Section 4 will briefly describe ANOVA models and their connection to regression. In section 5, we will introduce logistic regression, the method used when the response variable is a binary or yes/no variable. Finally, we will examine confounding and effect modification (or interaction), two important features of multivariate model building.

2. Simple Linear Regression

2.1. Introduction

Suppose we wish to be able to predict the total number of convalescence days in hernia repair patients. There may be a difference in what we should predict for convalescence days depending on the type of surgery, so we might want to compare the mean number of days in the two treatment groups. The mean number of convalescence days in the experimental group is 10.1 with standard deviation 7.6, and in the control group the mean number is 11.8 days with standard deviation 8.0. We can estimate the difference between the two groups as 1.7 days, with 95% confidence interval (-4.6,1.4). The confidence interval includes zero, so there is no strong evidence that the mean number of convalescence days in the two groups is different. Comparing two groups is easy enough, but what if we want to now look at the average numbers of convalescence days but using information about a continuous variable, like preoperative health profile score (based on the Nottingham health profile questionnaire[2])? We can't simply calculate mean values, because there are no natural groups into which we can divide the health score. We could divide the score into high and low values, but this would waste information about the differences that exist between subjects within the high and low groups. The first thing we might do is calculate the correlation coefficient between the health profile score and the number of convalescence days. This calculation gives a correlation of 0.240, with a 95% confidence interval (0.057,0.408) indicating that there is a statistically significant (but moderate in value) association between the two variables (the confidence interval does not include zero). However, we may want to know more about the relationship, for example by how much the response (convalescence days) varies as the predictor (health profile score) changes. The common approach to this problem is known as simple linear regression. Given a set of data where the number of convalescence days and the health profile score are measured on each patient in a group of patients, simple linear regression is a method for deriving and describing the straight line that best describes the relationship (if any) between the two variables. Consider Figure 5.1, a scatterplot with values of the response along the vertical axis and values of the predictor along the horizontal axis. It is clear that there is great variability in the convalescence days in these subjects, but that there may be a trend towards increased convalescence days with increased health profile score. We would like to fit the best straight line to this set of points, which would represent the trend, if any.

5

2.2. The Model

The simple linear regression model for this problem is given by:

$$y = \beta_0 + \beta_1 x$$

where x, the number of convalescence days, is to be predicted from x, the health profile score. The parameters β_0 and β_1 are regression coefficients that relate y to x. In particular, β_1 is the regression slope, so that a one unit change in x leads to a change of β_1 in y, and β_0 is the intercept, which is the predicted value of y when x = 0 . The basic assumption made is that the model is linear (that is, the relationship between convalescence days and health score is a straight line). Of course, it is extremely unlikely that this model will be exactly correct for all data points, in that we do not expect the values of x and y for all patients to lie exactly on a straight line, and we must account for this. We do so by adding a small amount of error, ε, to the model, so that the equation for the *i*th patient becomes

$$y_i = \beta_0 + \beta_1 x_i + \varepsilon_i. \quad (1)$$

The ε are often called "noise", residuals, or random errors. While we usually do not expect these errors to follow any predetermined pattern, we do make assumptions about their behavior. The standard set of assumptions is that they are:

1. independent from each other,
2. follow a Normal distribution with zero mean and
3. have a constant standard deviation σ throughout the range of the data.

Estimation of the coefficients β_0 and β_1 is done using the method of least squares. Briefly, this method chooses the estimated coefficients so as to minimize the sum of the squared differences between the estimated values and the actual values for . In other words, from equation (1), since we have $\varepsilon = y - \beta_0 + \beta_1 x$, we minimize the sum of the ε^2s. In our example, the value of β_1 represents the increase in the number of convalescence days y corresponding to an increase of 1 unit on the health profile score. The value of β_0 is the mean number of convalescence days for an individual with a health profile score of zero. Applying the least squares technique to the data in Table 5.1, we find that the "best" straight line regression model is:

$$y = 9.43 + 0.27x. \quad (2)$$

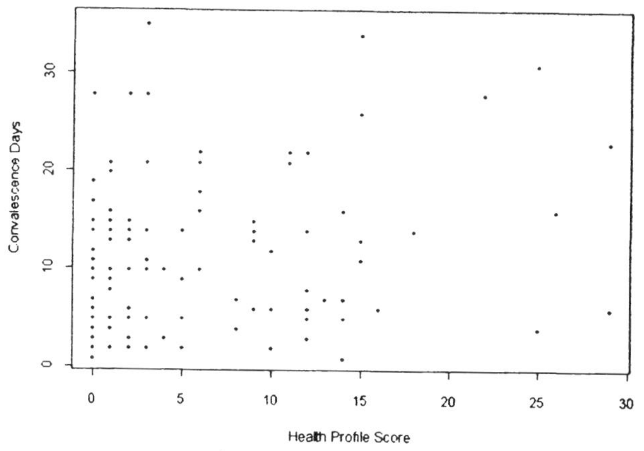

Fig. 5.1. Scatterplot of convalescence days vs. health profile score.

Therefore, the mean number of convalescence days is predicted to be 9.43 for people scoring 0 on the health profile index (i.e., with no major health problems) and this number increases by 0.27 days for every increase of 1 in the index. We refer to the estimates of the β parameters using the ^ character, so that we estimate $\hat{\beta}_1 = 0.27$ and $\hat{\beta}_0 = 9.43$.

Note that β_0 may not be clinically meaningful. It requires that the value $x = 0$ is meaningful. For example, if body temperature were being used as the x variable, it would not make any sense to think of a body temperature of zero, and then β_0 would have no direct physical interpretation. It would still be important, however, to estimate β_0, in finding the best fitting straight line to a set of data. This illustrates an important general principle: regression equations are only safely used over the range of the data they were calculated from. For example, a regression equation showing the decline of bone mineral density with age calculated on female subjects over age 50 would not necessarily be applicable for predicting bone density in young adult females or in males.

It is also useful to look at the estimate $\hat{\sigma}_r$, the standard deviation of the residuals. For this model, $\hat{\sigma}_r = 7.59$ indicating the size of the residual error standard deviation (measuring the unexplained variability in the data).

Often we want to compare mean values of a variable in two different groups. We can do this using model (1), with the predictor variable x taking on only two values, 0 in one group and 1 in the other. Such an x variable is referred to as a "dummy" variable. Then, the coefficient β_1 represents the difference between the mean level of the response in the two groups. In our example, we could set $x = 0$ in the standard treatment group and $x = 1$ in the experimental treatment group. Again applying least squares gives the model:

Table 5.1. Convalescence study data

Pt.	Convalescence Days	Health Score	CSST	Group	Occupn	Size	General	Sex	Smoking (years)
1	21	6	0	0	1	0	1	1	0.0
2	4	25	0	1	1	1	1	1	0.0
3	3	0	0	1	0	1	1	1	0.0
4	12	0	0	0	0	1	1	1	0.0
5	4	1	0	1	0	0	1	1	12.0
6	5	2	0	1	0	1	1	1	0.0
7	11	3	1	0	0	0	0	1	9.0
8	20	1	0	1	0	0	1	1	0.0
9	28	22	0	0	1	0	1	0	0.0
10	3	12	0	0	0	1	1	1	0.0
11	22	12	1	1	0	1	1	1	0.0
12	5	14	0	1	0	1	1	1	0.0
13	10	0	0	0	0	1	0	1	0.0
14	12	10	0	1	0	1	1	1	0.0
15	15	9	0	1	0	0	1	1	30.0
16	9	1	1	0	0	0	0	1	40.0
17	5	1	0	1	0	1	1	1	0.0
18	1	0	0	1	0	1	1	1	0.0
19	5	12	0	0	0	1	1	1	3.5
20	14	5	0	1	0	1	1	1	10.0
21	7	14	0	0	1	0	1	1	0.0
22	10	2	0	0	1	0	0	1	0.0
23	13	15	0	1	0	0	1	1	50.0
24	6	29	0	0	0	1	1	1	40.0
25	13	9	0	1	1	0	1	1	40.0
26	17	0	1	0	0	1	0	1	0.0
27	7	13	0	1	0	1	1	1	0.0
28	2	0	0	1	0	1	1	1	0.0
29	6	2	0	0	0	1	0	1	31.0
30	8	1	0	1	1	1	1	1	0.0
31	2	10	0	0	0	0	0	1	0.0
32	3	2	0	0	1	0	1	1	0.0
33	10	4	0	0	0	0	1	0	0.0
34	4	0	0	0	0	1	1	1	0.0
35	14	12	0	1	1	0	1	1	40.0
36	6	2	0	0	0	1	0	1	0.0
37	21	3	0	0	0	1	0	1	0.0
38	21	11	0	1	0	0	1	1	0.0
39	7	0	0	1	0	0	1	1	0.0
40	6	16	0	1	0	1	1	1	0.0
41	11	15	0	0	0	1	0	1	0.0
42	5	5	0	0	1	1	0	1	2.0
43	28	3	0	0	1	0	0	1	20.0
44	14	0	0	0	1	1	0	1	0.5
45	2	1	0	1	0	1	1	1	18.0
46	6	0	0	0	1	1	0	1	4.0
47	6	0	0	1	0	1	1	1	0.0
48	2	2	0	1	1	0	1	1	0.0

Table 5.1. (continued)

Pt.	Convalescence Days	Health Score	CSST	Group	Occupn	Size	General	Sex	Smoking (years)
49	6	0	0	0	0	1	1	1	0.0
50	10	4	0	1	1	0	1	1	30.0
51	6	12	0	0	0	1	0	1	3.5
52	19	0	1	1	1	1	1	1	0.0
53	2	3	0	0	0	1	0	1	0.0
54	28	2	0	0	0	0	0	0	0.0
55	10	3	0	0	0	1	0	1	40.0
56	3	2	0	1	0	1	1	0	0.0
57	16	14	1	0	0	0	1	1	12.0
58	13	1	0	1	1	0	1	1	0.0
59	10	1	0	1	1	0	1	1	0.0
60	35	3	1	0	1	0	1	1	0.0
61	3	4	0	0	0	0	0	1	45.0
62	16	6	1	1	1	0	1	1	0.0
63	18	6	0	0	0	0	0	1	0.0
64	2	1	0	0	1	0	0	1	0.0
65	5	0	0	1	0	1	1	1	10.0
66	15	0	0	0	1	1	0	1	0
67	22	11	1	0	0	0	0	1	0
68	28	0	0	0	0	0	0	1	42
69	6	10	0	1	1	0	1	1	0
70	1	14	0	1	0	1	1	1	50
71	6	9	0	0	0	1	0	1	0
72	11	0	1	1	1	0	1	1	25
73	31	25	1	1	1	0	1	1	0
74	8	12	0	0	1	0	0	1	0
75	2	5	0	0	1	0	0	1	0
76	5	2	0	1	1	1	1	1	0
77	10	0	1	0	0	0	1	1	5
78	10	3	0	1	0	0	1	1	0
79	2	0	0	1	0	0	1	1	0
80	16	26	0	1	0	1	1	1	4
81	14	1	0	0	0	0	1	1	0
82	26	15	0	0	0	0	0	0	0
83	13	2	0	0	0	1	1	1	40
84	15	0	0	0	0	0	0	1	0
85	5	3	0	1	0	0	1	1	0
86	34	15	1	1	1	0	1	1	37
87	10	1	0	0	0	0	0	1	0
88	4	8	0	1	0	0	1	1	0
89	14	18	0	0	0	1	0	1	11
90	14	0	0	1	1	1	1	1	0
91	14	9	0	1	1	1	1	0	0
92	22	6	0	1	0	0	1	1	0
93	15	2	0	0	0	1	1	1	10
94	15	1	0	0	0	1	1	1	0
95	9	5	0	0	0	1	1	1	0
96	9	5	0	0	0	1	1	1	0

Table 5.1. (continued)

Pt .	Convalescence Days	Health Score	CSST	Group	Occupn	Size	General	Sex	Smoking (years)
97	5	0	0	0	0	1	1	1	0
98	5	2	0	1	0	1	1	1	0
99	14	2	1	0	0	1	0	1	0
100	10	2	0	1	1	1	1	1	0
101	2	0	0	1	0	1	1	1	0
102	23	29	0	0	0	1	1	1	0
103	10	6	0	1	1	0	1	1	40
104	1	0	0	1	0	1	1	1	0
105	15	1	1	1	0	1	1	1	0
106	9	0	0	0	0	1	1	1	15
107	7	8	0	0	0	1	1	1	0
108	1	0	0	1	0	0	1	1	0
109	1	0	0	0	0	1	0	1	1
110	16	1	1	1	0	1	1	1	0
111	21	1	1	1	0	0	1	1	20
112	14	3	0	0	0	1	0	1	45

$$y = 11.81 - 1.75x.$$

The interpretation of the coefficients in this model is convenient. The coefficient β_0 refers to the mean number of convalescence days in the group with $x = 0$, which is the standard treatment group. Hence, the mean number of convalescence days in this group is 11.81. The parameter β_1 refers to the difference in mean level between the two groups, meaning that the average number of convalescence days in the experimental treatment group is 11.81-1.75 = 10.06. In other words, the new surgical technique leads to a $\hat{\beta}_1$= -1.75 decrease in convalescence days on average. Note that the two means of 11.81 and 10.06 for the conventional and laparoscopy groups respectively are equal to the observed mean values in the data for these two groups given in section 2.1. Here, $\hat{\sigma}_e = 7.77$, indicating that the residual standard error is roughly similar to that from our first model.

2.3. Statistical Inference About the Regression Coefficients

Standard statistical procedures can be used to derive confidence intervals for β_0 or β_1 and to test whether the predictor variable is useful in explaining the variability in the response. Recall that $\hat{\beta}_1$ is the estimated value of β_1 and define the estimated standard error of $\hat{\beta}_1$ to be

$$SE\left(\hat{\beta}_1\right) = \frac{s}{\sqrt{\left(x_1 - x -\right)^2}}$$

where

$$s = \sqrt{\frac{e^2}{n-2}},$$

is the number of observations and the e are the observed residuals ($e = y - \beta_0 - \beta_1 x$). The quantity $\hat{\beta}_1/SE(\hat{\beta}_1)$, has the t-distribution with n-2 degrees of freedom, where n is the sample size. This can be used to test the null hypothesis $H_0{:}\beta_1 = 0$, which, if rejected, suggests that β_1 is useful in predicting the response. A 100(1-α)% confidence interval can be derived from the formula:

$$\hat{\beta}_1 \pm t_{\alpha/2,\, n-2} SE\left(\hat{\beta}_1\right),$$

where $t_{\alpha/2,n-2}$ is the t-distribution $\alpha/2$ critical value for n-2 degrees of freedom. If this confidence interval does not include 0, this is evidence that x is an important predictor. In our case, the 95% confidence interval for β_1 in the health profile score is (0.07, 0.48), indicating that plausible values for the increase in convalescence days with an increase of one unit on the health scores range from 0.07 to 0.48. So our results are consistent with almost no effect (0.07 days) but also with a moderate effect (almost half a day of convalescence per unit increase in the score).

We can also derive confidence limits for the coefficients for the model comparing the experimental treatment group to the standard treatment group. The coefficient is –1.75, with standard error 1.48, so a 95% confidence interval is given by (–4.65,1.15). It is worth noting that the test that this coefficient is equal to zero and the associated confidence interval are identical to that which would be derived from the t-test under the assumption of equal variances in the two groups. We will see how this generalizes to the relationship between multiple regression and analysis of variance in Section 3.

2.4. Checking the Model

When using regression models, it is important to be aware of and understand the underlying assumptions of the model. First, the model (1) is linear – a unit increase in the health profile score corresponds to an increase of β_1 in convalescence days for any value of the health score. It is conceivable that this assumption is not correct, that the rate of increase in convalescence days is higher for higher health profile scores compared to lower health profile scores, for example. To investigate this assumption, the scatterplot with the y variable along the vertical axis and the variable along the horizontal axis can help (see Fig. 5.1). If a curved or other pattern is observed here, steps can be taken to address it, usually by transforming the x or y variable. For example, rather than using x, we might try x^2, or rather than y we might try log(y).

Another essential part of the model checking process is to look at the residuals. Recall that the model virtually never fits perfectly and we assume that the true model fits with a small amount of error added, as in equation (1). We cannot observe the actual values of these errors, but we can get an idea of what they look like by comparing the observed values of the response to the values that would be predicted by the model. For example, from Table 5.1 we see that patient 5 had a health profile score of 1, so that equation (2) implies that his or her predicted number of convalescence days is 9.43 + 0.27*(1) = 9.70. The true number of convalescence days was 4, meaning that the residual for this patient was 9.7 - 4 = 5.7.

Knowing that we expect the true errors to be normally distributed with the same standard deviation, we can check the residuals to see if they roughly fit this distributional assumption. If so, the model fits the data well and that's good. If not, we need to investigate the model in more detail. One way to check this assumption is to plot the residuals against x. Figure 5.2 is a plot of the residuals from model (1) plotted against the health profile scores. There appears to be a reasonable scatter of the residuals, although higher residuals are more prominent for lower values of the profile score. Figure 5.3 is a histogram of the residuals, which shows that the distribution is somewhat right-skewed and does not look approximately normal. If this is the case, what can be done? Several possibilities exist, but the most common approach is to transform either the response or the predictors to achieve a model with better residual properties. A thorough discussion of the options for transformations can be found in Neter et al.[3]

3. Multiple Linear Regression

3.1. Introduction

What if we have many predictor variables, each of which has an independent influence on the outcome? Perhaps we would like to examine the association between total convalescence days and the health profile score and group together. One possible approach would be to try several simple linear regression models, one for each of our possible predictor variables. This will give information about their individual effects on the total number of convalescence days but not on the joint effect of all variables together, or on the effect of one of the variables adjusted for the other. We can use multiple regression, which simultaneously associates many predictors to a continuous response, to accomplish both these tasks.

3.2. The Model

A multiple regression model with two predictor variables is written as:

$$y = \beta_0 + \beta_1 x_1 + \beta_2 x_2 + \varepsilon \qquad (3)$$

The model works in essentially the same way as the simple regression model (1) but interpretation of the coefficients changes slightly because of the presence of other variables in the model. Now, β_1 refers to the change in y associated with a unit change in x_1 after adjusting for the effect of x_2. In other words, for a fixed level of x_2, the mean change in y when x_1 increases by 1 is β_1. The analogous interpretation holds for β_2 when x_1 is held constant. These definitions are unchanged when x_1 is a group variable, for example the dichotomous ($x = 0$ or $x = 1$) variable we saw in section 2.2 representing the type of hernia repair received. Here, β_1 would refer to the average difference between y values for two people with identical levels of x_2 but in the two different treatment groups.

In our example, suppose we want to look at the effects of the health profile score and group together. Let x_1 denote the health profile score and x_2 be 0 in the standard treatment group and 1 in the experimental group. The fitted model is:

$$y = 10.29 + 0.27x_1 - 1.79x_2$$

meaning that for each unit increase in health profile score the mean increase in convalescence days is 0.27, assuming that the group is fixed—that the patients being compared are in the same treatment group. Similarly, if we compare two people with the same health profile score but in the two different treatment groups we would expect that the patient in the experimental group would have 1.79 fewer days

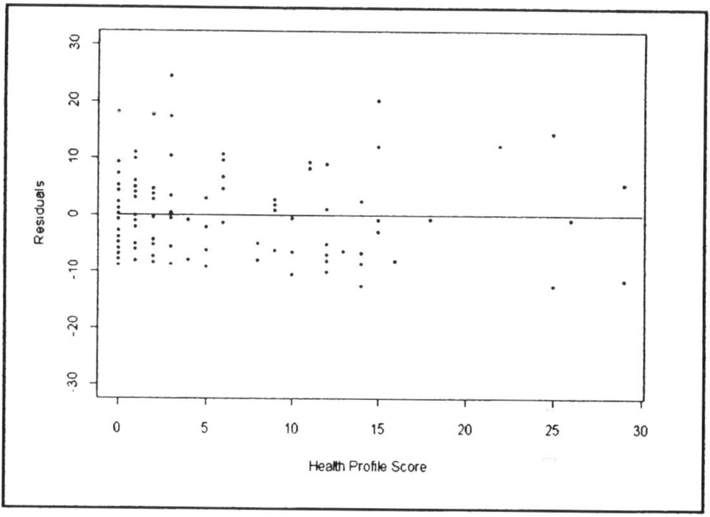

Fig. 5.2. Scatterplot of residuals from best model vs. health profile score.

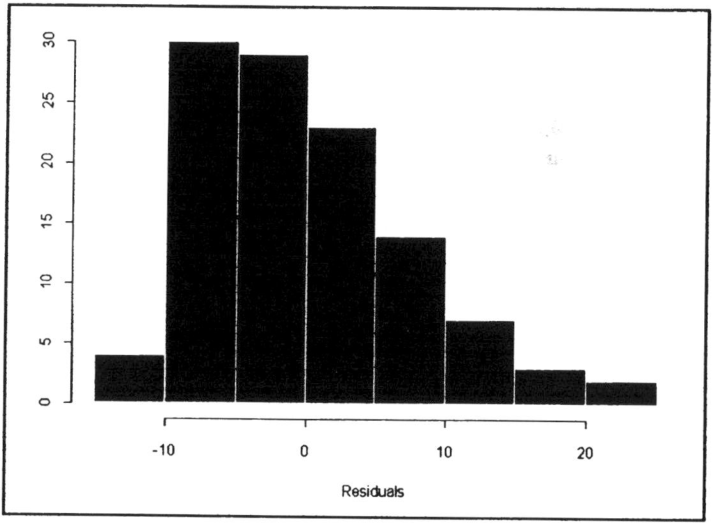

Fig. 5.3. Histogram of residuals from last model.

of convalescence. It is interesting to note that these coefficients are very close to those in the original univariate models. However, the standard error of the residuals (7.53) has declined relative to both of the univariate models.

Many readers will be familiar with analysis of variance or ANOVA techniques. ANOVA generalizes the t-test to comparisons of more than 2 groups. ANOVA and multiple regression are closely related. Since there are more than two groups being compared, we have to look at more than just a single mean difference. The method for testing the whether the mean level in all of the groups is the same follows a general pattern similar to that for the t-test.

To illustrate, consider a one-way ANOVA with three groups. The standard ANOVA model for this problem is

$$\mu_i = \mu + \alpha_i \qquad (4)$$

where μ_i, the mean for the *i*th group, $i = 1,2,3$, is equal to a grand mean μ plus an effect for the group, α_i. Conventionally, we set the sum of the α_i to be zero, to allow us to fit the model. The regression model:

$$y = \beta_0 + \beta_1 x_1 + \beta_2 x_2 + \varepsilon$$

with $x_1 = 1$ in group 1, 0 otherwise, and $x_2 = 1$ for subjects in group 2, and 0 otherwise, can be shown to be equivalent to the ANOVA model (4) above. A small amount of algebra can be used to show that if $\beta_0 = \mu$, $\beta_1 = \alpha_1 - \alpha_3$ and $\beta_2 = \alpha_2 - \alpha_3$, the models are equivalent.

Recall that the t-test can be formulated as a simple linear regression model. We have described one example above and hinted at the general principle that ANOVA, and more complicated models like analysis of covariance (ANCOVA) can be thought of as multiple regression models with appropriately defined coefficients. Further discussion of this can be found in the book by Neter et al.[3]

3.3 Model Selection

To model convalescence days, we have several options. Considering the variables in Table 5.1, there are $2^7=128$ possible combinations of the seven independent variables that could be entered into a model. How should we decide how many and which independent variables we should include in the model? Having too few independent variables will result in a model with sub-optimal predictions, while having too many adds unwanted variability to the estimation process. When faced with multiple possible predictors, how do we decide a) which to consider as part of a model and b) which model is the best available, given the data and our knowledge of the problem? This is a complicated process, and there are many factors that influence the way in which it is done.

Many researchers use model selection techniques, such as backwards or forwards stepwise regression which systematically builds a model based on a variety of criteria. However, it is often possible that models that have good statistical properties are lacking in real clinical properties, in that variables that are in the model may be of less interest to research than variables that are out of the model. Further, stepwise model selection with a large number of predictors can often give results that are statistically significant purely by chance. Stepwise procedures decide between two models using one of several possible criteria. Some of the most common are the Akaike information criterion and the Bayes information criterion.[4] All of these (and others) evaluate models based on criteria such as fit (better fitting models are better) and complexity (in general, if two models fit the data equivalently well, the less-complex model is better).

So while stepwise regression can be useful, it has to be used with extreme caution to be sure that the results make sense from the clinical and research perspective. It is better to have a good understanding of the variables in the dataset. Ideally, descriptive statistics and clinical knowledge will indicate an order in which to consider variables in the model. For example, in nonrandomized studies it is customary to use age and sex of patients in the model even when they are weak predictors of the response, just because it is appropriate to discuss the models by comparing the results for patients of the same sex and age.

Continuing with our example, suppose we want to try to select the most appropriate model for our response using all of the potential predictor variables in Table 5.1. The best model using the AIC criterion is given by:

$$y = 17.74 + 0.24x_1 + 8.98x_2 - 1.92x_3 - 3.34x_4 - 7.11x_5$$

where x_1 is the health profile score, $x_2 = 1$ if the patient was compensated by the government and 0 if not, $x_3 = 1$ if the patient was in the experimental group, $x_4 = 1$ if the hernia was large, and $x_5 = 1$ if the patient was female. In this case, the model is relatively satisfactory from a clinical point of view because it contains most of the terms that we would consider important, and the variables that were not included (smoking, occupation, and use of general anesthetic) were not a priori thought to be particularly important. From a statistical point of view, this model is an improvement over other simpler ones. The estimated standard deviation of the residuals, $\hat{\sigma}_r$, is smaller when the model is better, that is to say there is less unexplained variation in the data. In this case, the simple model with group alone had $\hat{\sigma}_r = 7.77$ while this model has $\hat{\sigma}_r = 6.29$. Finally, it is reassuring to note that the coefficient for group is −1.92, very similar to the −1.75 found in the simple model (and using the t-test). This suggests that even after adjustment for the other important predictors of total convalescence days, there is a persistent effect of treatment group. The experimental group had, on average, slightly less than two fewer convalescence days.

3.4. Statistical Inference for Multiple Regression Models

Tests and confidence intervals for the individual β coefficients in the multiple regression model operate in much the same way as they do in the simple linear regression model. The quantities $\hat{\beta}_1/SE(\hat{\beta}_1)$ have t-distributions with n-p-1 degrees of freedom, where n is the number of subjects in the dataset and p is the number of parameters (or β coefficients) in the model. This can be used to test $H_0{:}\beta_i = 0$, which works in an analogous way to the simple case, with one important difference. The test only applies in the context of the current model. If a variable makes a statistically significant contribution to a particular model, that conclusion cannot necessarily be generalized to mean that the variable is important in all situations. In other words, if the model is changed such that different variables are included, the apparent effect of the variable may change. See section 5 below. 100(1-α)% confidence intervals can be derived from the formula:

$$\hat{\beta}_i \pm t_{\alpha/2,\,n-p-1} SE\left(\hat{\beta}_i\right)$$

where $t_{\alpha/2,n-p-1}$ is the t-distribution $\alpha/2$ critical value for n-p-1 degrees of freedom. Most software packages calculate both $\hat{\beta}_i$ and $SE(\hat{\beta}_i)$, and many will automatically derive the confidence intervals.

3.5. Checking Model Assumptions

Diagnostic tests to check model assumptions for multiple regression models operate in a way completely analogous to simple regression. Checking linearity using scatterplots of the response against the various predictors and plots of the residuals are the two main methods for determining whether the model is appropriate. For certain problems, there are simple solutions to the violation of assumptions. If the pattern of the residuals appears curved, a polynomial model (using higher powers of the predictors) may be appropriate. If the variance of the residuals appears unequal across the range, a weighted regression approach may help.

4. Logistic Regression

How could we approach the convalescence example if we were interested in predicting whether or not patients had a convalescence of more or less than one week? This outcome would typically be coded as 1 if the convalescence was longer than a week, and 0 if shorter. We would not be able to apply linear regression here, mainly because the outcome is not continuous – it can only be 0 or 1. This would lead to residuals that are not normally distributed, and violation of the linear regression assumptions.

Many medical outcomes do not satisfy the assumptions listed in the previous section, in particular that the outcome be measured on a continuous scale. Outcomes such as alive/dead, on/off or success/failure all have a similar binary structure that does not allow us to use linear regression to analyze them. Statistical techniques that take these problems into account are available. Logistic regression, which models the probability of one of the two outcomes, is a convenient method of analyzing such data.

Logistic regression is an extension of linear regression. In logistic regression, the outcome is no longer a continuous variable but a dichotomous variable, and we are interested in estimating the probability that the outcome will take on one of its two values. Consider an example where we want to associate death during surgery to various factors. We would then be interested in predicting the probability of death as it relates to these factors. Linear regression can be used in some cases for these data, but it can lead to serious problems such as predictions that fall outside of the [0,1] range. A mathematical rearrangement of the linear regression model is necessary so that we can have a model in a more useful form. This also means that the β parameters in logistic regression have different interpretations than do the ones in ordinary linear regression. In logistic regression, e^{β_1} refers to the odds ratio of the outcome (e.g., death, or convalescence of more than seven days) for two subjects that differ by one unit of x_1. The odds ratio is similar to but somewhat different from the more commonly understood relative risk or risk ratio. The relative risk for comparing the outcome in one group to that in another group is simply:

$$RR = \frac{Risk_1}{Risk_2}$$

the ratio of the risks, or probabilities of the outcome, in the two groups. The odds ratio is slightly different,

$$OR = \frac{Odds_1}{Odds_2}$$

where the odds of the disease in each group is given by

$$Odds = \frac{Risk_i}{1 - Risk_i}$$

For example, a risk of disease of 0.5 (50%) corresponds to an odds of disease of 1.0, while a risk of 0.2 corresponds to an odds of 0.2/0.8=0.25.

Usually, the relative risk is the parameter that we would like to draw inferences about, so why do we use the odds ratio? The main reason is mathematical convenience; logistic regression easily produces estimates of the odds ratio. Further, in case-control studies, the odds ratio is the only parameter comparing the two groups that we can estimate. Finally, when the disease or outcome being analyzed is rare, the odds ratio is nearly the same as the relative risk. This happens because when the outcome is rare, 1-$Risk_i$ is approximately 1, so $OR \cup RR$ in the above equations. For example, if the risk in one group is 0.1% and in the other 0.2%, the risk ratio or relative risk is 2, while the odds ratio is $(^{0.02}/_{0.98})(_{0.01}/_{0.99})$=2.02, virtually indistinguishable.

Formally, writing the risk for an event as the probability p that it occurs, the model for a logistic regression with two predictors is

$$\log it(p) = \log\left(\frac{p}{1-p}\right) = \beta_0 + \beta_1 x_1 + \beta_2 x_2 \qquad (5)$$

meaning that we are modeling the natural logarithm of the odds rather than the actual probability. This log odds is often referred to as logit(p), hence the name logistic regression. This transformation does not cause much of a problem, because we can easily convert back to probabilities since:

$$\log it(p) = \log\left(\frac{p}{1-p}\right) = \beta_0 + \beta_1 x_1 + \beta_2 x_2 \Rightarrow p = \frac{e^{\beta_0 + \beta_1 x_1 + \beta_2 x_2}}{1 - e^{\beta_0 + \beta_1 x_1 + \beta_2 x_2}}$$

Consider again the example of convalescence days after hernia repair. We might want to examine the probability that a patient convalesces for more than a week (i.e., the number of convalescence days is greater than 7), using group and size of the hernia (large vs. small) as potential predictors. We would then develop the model (5) and the coefficients would be interpreted as stated above. Fitting the model with x_1 being coded as 1 for a large and 0 for a small hernia and x_1 defined as 0 for subjects from the standard treatment group and 1 in the experimental group,

$$\log\left(\frac{p}{1-p}\right) = 1.24 - 1.11x_1 - 0.56x_2 \,.$$

Interpreting this model is simple. To compare the risk of convalescence of greater than a week in the two treatment groups, the value e^{β_2} or in this case $e^{0.56} = 0.57$ is the odds ratio for comparing these treatments. This means that the odds of convalescing for more than a week are 0.57 times as large in the experimental group as they are in the treated group.

Confidence intervals and tests for the parameters of these models are constructed in the same way as in the linear regression model. For example, to get lower and upper limits of a 95% confidence interval for β_1, we calculate 1.11-1.96*$SE(\hat{\beta}_1)$ and 1.11 + 1.96*$SE(\hat{\beta}_1)$, respectively, where $SE(\hat{\beta}_1)$ is the standard error of the estimate for β_1, given by all computer packages that handle logistic regression. Further details about logistic regression can be found in the workbook text by Kleinbaum[5] or the textbook by Hosmer and Lemeshow.[6]

5. Confounding and Causation

In multiple regression, it might arise that as well as being associated with the response, two predictor variables are associated with each other. If this happens, the coefficient for one of the variables can be distorted by the absence of the other variable in the model, and vice versa. Statisticians and epidemiologists call this confounding, and it typically occurs when two predictors that are strongly related, such as smoking and alcohol consumption, are both of interest in a model. At its worst, confounding can lead to completely spurious results. For example, high water consumption is associated with low blood pressure. Water consumption is higher in people who exercise frequently, and people who exercise frequently have lower blood pressure, while water consumption, after taking into account exercise frequency, has little or no effect on blood pressure. We say that the relationship between water consumption and blood pressure is confounded by exercise level.

Consider the model given in Table 5.2 for our sample dataset. This model does not contain the CSST variable. Note that the coefficient for sex is -5.42 and has wide confidence interval (-11.41,0.58) which includes the value 0. Addition of the CSST variable to the model changes the coefficient of sex to -7.12 and the confidence interval to (-12.53,-1.70) which is narrower and does not include the value 0. This indicates that there is confounding between these two variables, i.e., that the CSST compensation is associated with both sex and the total number of convalescence days.

This leads to the broader question: does linear regression tell us anything about causation? The only thing that linear or logistic regression really tells us is that there exists a relationship between two or more variables. Causation is not necessarily implied, because, for example, the true relationship could be confounded by some third variable that we did not measure, or the causal relationship could actually go the other way, with the outcome causing the predictor variable. More evidence than just a regression relationship is needed to justify a causal relationship, such as biological plausibility, and the elimination of other possible causes such as ecological fallacy. See Rothman and Greenland[7] for further discussion of causation.

6. Effect Modification or Interaction

A second problem arises when two variables are associated such that the level of one of the variables modifies the effect of the second variable on the response, or the effect of the second variable on the response is different depending on the level of the first variable. We call this effect modification or interaction and account for it

Table 5.2. Results of model with sex removed

	Coef.	Std. error	95% CI
Intercept	18.06	3.13	—
HPS	0.24	0.10	(0.04, 0.44)
Group	-1.75	1.36	(-6.25, 0.92)
Size	-4.52	1.37	(-7.21, -1.83)
Sex	-5.42	3.07	(-11.41, 0.58)

Note: HPS=Health profile score, Group=1 if patient was in the experimental group, Size=1 if the hernia was large, and Sex=1 if female

essentially by building separate models or including interaction terms in a single model to account for the association between the second variable and the response for different levels of the first variable. For example, the effect of exercise on heart attack risk may be different in smokers than in nonsmokers.

How is this modeled? A multiplicative effect modification or interaction is addressed in the model using the product of the two interacting variables. In the model:

$$y = \beta_0 + \beta_1 x_1 + \beta_2 x_2,$$

for example, if we want to look at the interaction between x_1 and x_2, we add a term of the form $x_1{}^*x_2$ so that the model becomes:

$$y = b_0 + b_1 x_1 + b_2 x_2 + b_{12} x_1{}^* x_2 \qquad (6)$$

This means that the increase in y corresponding to a unit increase in x_1 now depends on x_2 and is given by

$$\beta_1 + \beta_{12} x_2$$

Let's consider a model with interaction in the hernia data. Using the health profile score and government compensation (CSST) as predictor variables, we could fit model (1) with the following results:

$$y = 8.03 + 0.27x_1 + 9.21x_2$$

where x_1 is the health score and x_2 is 1 if compensation was received and 0 otherwise. The coefficient for the health profile score is interpreted as before: for one unit increase in score, the total convalescence days increase by 0.27. The interaction model (3) gives the following results:

$$y = 8.43 + 0.20x_1 + 6.64x_2 + 0.46x_1{}^*x_2,$$

so the increase in convalescence days for a unit increase in health profile score is:

$$0.20 + 0.46x_2.$$

So, for people who received compensation ($x_2 = 1$) the increase in convalescence days is 0.66 for each unit increase in health profile score, while for patients who did not receive compensation, the mean increase was only 0.20 days. The interaction suggests that the combination of poor health and compensation is associated with longer recoveries than does the model without interaction.

It is usually important to look for interaction. In large models, however, there can be a huge number of potential interactions (we could, in principle consider products of three or more terms too, and in fact in a model with five variables, there are 2^5 or 32 possible interactions!). For this reason, it is important to try to prespecify interesting interactions to look for in the models. Prespecification can be done on the basis of clinical reasoning or exploratory statistical analysis that may demonstrate which potential interactions look interesting.

7. Conclusion and Literature Review

5

In this Chapter we have introduced two primary methods for analysis of multivariate data, linear and logistic regression. These methods, in a similar framework, allow the analysis and prediction of continuous and binary outcomes, respectively, as a function of other variables.

Other models that accommodate more complex structures are also available. When multiple correlated outcomes are of interest (for example, repeated measurements on the same patient), these correlations must be taken into account. Various methods for both continuous and binary outcomes, such as generalized estimating equations and random effects models provide estimates of relationships which consider the correlations between outcomes.[8] The models we have looked at here can also be generalized to other types of outcomes, such Poisson regression for counts, so that appropriate inferences can be made when these types of variables are of interest.[9]

Computer programs to perform the analyses in this Chapter are readily available in a bewildering variety of statistical packages of varying quality and ease of use. Major packages include SAS,[10] SPSS,[11] S-Plus,[12] and STATA.[13] There are hundreds of texts on the topics discussed in this Chapter, a few of which are listed in the references. For linear regression, the book by Neter et al[3] is a comprehensive look at design and analysis of regression studies. For a perspective on linear regression aimed more at the medical sciences, Selvin[14] is comprehensive and easy to read. The workbook text by Kleinbaum[5] provides a systematic introduction to logistic regression, with worked examples and step by step procedures for modeling the odds ratio. The book by Hosmer and Lemeshow[6] is a more detailed work primarily designed for epidemiologists and biostatisticians.

Selected Readings

1. Rubin D. Multiple Imputation for Non-Response in Surveys. New York: Wiley 1996.
2. Hunt SM, McEwen J, McKenna SP. Measuring health stats: A new tool for clinicians and epidemiologists. J Royal Coll Gen Pract 1985; 35:185-88
3. Neter J, Kutner M et al. Applied Linear Statistical Models. Fourth Edition. Chicago: Irwin 1996.
4. Kass RE, Raftery AE. Bayes Factors. J Am Stat Assoc 1995; 90:773-795.
5. Kleinbaum DG. Logistic regression: A self learning text. New York: Springer-Verlag, 1984.
6. Hosmer DW, Lemeshow S. Applied Logistic Regression, John Wiley and Sons, New York, 1989.

7. Rothman K, Greenland S. Modern Epidemiology. New York: Lippincott-Raven 1997.
8. Diggle PJ, Liang K-Y, Zeger SL. Analysis of Longitudinal Data. Oxford: Oxford Science Publications, 1994.
9. McCullagh P, Nelder JA. Generalized Linear Models, 2nd Ed. London: Chapman and Hall 1989.
10. SAS Version 6.12. Cary, NC: The SAS Institute 1997.
11. SPSS Version 8.0. Chicago, IL: SPSS Inc 1997.
12. S-Plus Version 4.5. Seattle, WA: Mathsoft Inc. 1998.
13. STATA Statistical Software Release 5.0, Stata Corp, College Station, TX, 1997.
14. Selvin S. Practical Biostatistical Methods. Belmont, CA: Wadsworth 1995.

5

CHAPTER 6

Survival Analysis

David B. Wolfson and Cyr Emile M'Lan

1. Introduction

In this Chapter we shall attempt to explain some of the basic notions of the area of statistics known as survival analysis. In order to maintain continuity of the ideas introduced, perhaps contrary to convention, the illustrative example on survival of patients with malignant brain tumors has been placed at the end of the Chapter. This also helps ensure that the example is seen as it might be, in the real world, as a unit rather than consisting of a set of separate analyses. This topic certainly does not deal exclusively with "survival", that is, with an outcome of "death". Its use would be rather restricted if this were the case, whereas in surgery and other research, survival analysis is of crucial importance because its methods apply to a broad range of outcomes, including "time to remission", "length of postoperative hospital stay", "time to recurrence of disease", "time free of pain", "time to failure (of a prosthetic device)" and of course, "time to death".

If you were to peruse a volume of say, "The Archives of Surgery", you would almost surely encounter terms such as Kaplan-Meier estimates, log rank tests, and Cox's proportional hazards model. For example Vol 131, Jan.-June 1996, of The Annals of Surgery contains articles on "Multimodel-Therapy Breast Salvage in the Urban Poor with Locally Advanced Cancer" (Boddie et al), "Patient Selection for Hepatic Resection of Colorectal Metastases" (Wanebo et al), "Hepatitis C Viral Infection in Liver Transplantation" (Johnson et al), and "Surgical Aspects of Patients with Adenocarcinoma of the Stomach Operated on for Cure" (Soreide et al), whose statistical analyses include methods of survival analysis.

There is an abbreviated bibliography on survival analysis at the end of this Chapter. Some of these references contain histories of the development of the subject. We have therefore refrained from presenting this history, although it is worth mentioning one paper which shows that the subject had its birth at least three hundred years ago. In 1693 Mr. E. Halley, of Halley's comet fame, published an article entitled "An estimate of the degrees of mortality of mankind drawn from curious tables of the births and funerals at the city of Breslaw" in the Philosophical Proceedings of the Royal Society of London. Basic methods for computing life expectancies were proposed for the purpose of assessing life insurance premiums; these ideas have evolved into the rich and powerful methodology included by the vast subject of event history analysis. Let's plunge in and swim a few beginners' strokes.

Surgical Arithmetic: Epidemiological, Statistical and Outcome-Based Approach to Surgical Practice, by Lawrence Rosenberg, Lawrence Joseph, and Alan Barkun. ©2000 Landes Bioscience.

2. What Is Survival Analysis?

In its simplest form survival analysis is concerned with statistical inference about the probability distribution of the time from some well defined origin until some well defined event.

For example, we may wish to make a statistical statement about the distribution of time to failure of a prosthetic hip following hip replacement surgery. Our data might consist of the recorded dates of surgery of a group of patients together with the dates on which their prostheses were assessed to have failed. It could easily happen, though, that at the end of the study some of the patients have been lost-to-follow-up, their prostheses having not yet failed when the subjects were last seen. In addition, some patients, particularly those who have their surgery towards the end of the study, may still have intact prostheses. Further, it is even possible that some subjects may have been removed from the study because they began to experience pain. This type of data are depicted in Figure 6.1, assuming, for simplicity that the calendar dates of surgery are unimportant. Therefore, all subjects have been given a common origin even though they may have entered the study on different dates.

6

If a subject's prosthesis has failed by the end of the study, we denote the time of failure by x.

If a subject's failure time has not been observed by the end of the study (for the reasons mentioned above, or perhaps for other reasons) we say the failure time has been censored and denote the instant of censoring by a circle.

For example, subject 1 has experienced failure at about 9 years while subject 3's prosthesis is still intact at 10 years.

The distinguishing feature of survival analysis is that its methods allow for incompletely observed failure times. By far the most important source of incompletely observed failure times in survival analysis is censoring.

It is clear that were we to omit those observations which are censored, we would be discarding valuable information. For example, it is certainly informative to know that subject 3's prosthesis lasted longer than 10 years. In fact, it is intuitively obvious that if we were to obtain the times of only those where prostheses had failed, our sample would represent prostheses with short failure times.

Since censoring is central to survival analysis it is important to discuss some of its features.

3. Censoring

Some simple notation will help clarify our discussion. Henceforth we shall refer to the endpoint of interest as the failure time or survival time, although as was indicated in the introduction, it is not at all necessary that failure in the usual sense, be understood. It may be useful to think of the hip replacement example as we proceed.

We denote by T_i the true failure time for subject *i*. Subject *i*'s hip prosthesis has an intrinsic random failure time at the time of surgery.

We denote by C_i the censoring time for subject *i*. At the time of surgery subject *i* has the potential to be censored at some time C_i, which may be random or a constant. If, as is often the case in clinical trials such as this, the main source of

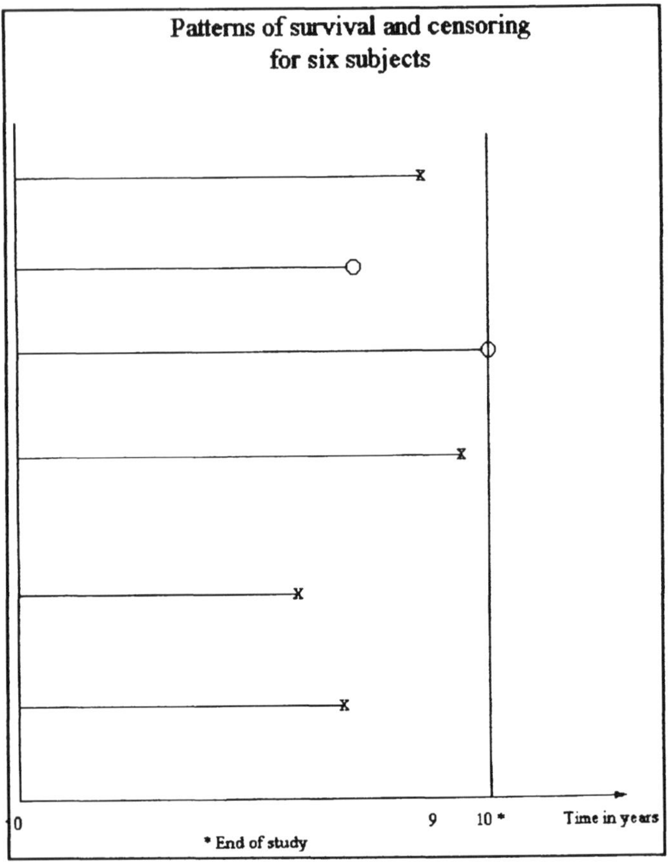

Fig. 6.1. Time to failure of a prosthetic hip following hip replacement surgery. (x) denotes failure and (o) denotes censored.

censoring occurs at the end of the study, each subject would have a known fixed potential censoring time (possibly unrealized), if we regard the entry times as fixed.

At the conclusion of a study the event time for subject i is denoted by X_i, where

$$X_i = \begin{cases} T_i \text{ if } T_i < C_i, \text{ and} \\ C_i \text{ if } T_i > C_i \end{cases}$$

In addition to observing X_i, we assume that we know whether or not the subject's failure time has been censored. The variable, δ_i, defined by:

$$\delta_i = \begin{cases} 1 \text{ if } T_i < C_i, \text{true failure has occurred} \\ 0 \text{ if } T_i > C_i, \text{ censored failure has occurred} \end{cases}$$

provides this information, and is appended to X_i as an indicator of whether a failure or a censoring has occurred. The complete data, ignoring covariates for the moment (see Section 8), may therefore be described by the pairs $(X_1,\delta_1),(X_2,\delta_2),\cdots,(X_n,\delta_n)$.

From Figure 6.1 the data for subject 1 would consist of the pair (9,1) and for subject 3, (10,0).

Now, certain types of censoring mechanisms are permissible. The strongest assumption that can be made concerning the censoring mechanism is that the random variables T_i and C_i are independent. That is, it is assumed that the failure occurs independently of censoring. This is always the case if the potential censoring time C_i is a constant, but more general situations might lead to independence between T and C; subjects who are lost to follow-up for reasons unrelated to (potential) failure of their hip prostheses fall into this category. This assumption, sometimes called random censoring, allows us to use all the standard methods of survival analysis.

In order to understand the meaning of random censoring it is worthwhile to consider an example in which this assumption would NOT hold. Suppose that hip replacement patients were removed from the study when they began to experience hip pain. Then, since it is possible that hip pain is a predictor of imminent prosthesis failure, the time to failure and time to removal from the study (censoring) would not be independent. By removing these subjects we bias the remaining survival times towards larger values.

6

4. Notation and Terminology

It is assumed that the pairs, (X_i,δ_i), $i = 1,2,\cdots,n$ are independent random variables. This would not be the case, for example, if early failure of a hip prosthesis in certain subjects led to late failure in other subjects. This might arise if some change in patient management occurs as a result of the failures observed early on in the study.

Interest in survival analysis, and in particular in much of surgery research, concentrates on inference about the probability distribution of the failure times T_i. This distribution is specified through the survivor function, defined as:

$$\begin{aligned} S(t) &= P(T \geq t), \text{ for all } t > 0 \\ &= 1 \text{ for } t \leq 0 \end{aligned}$$

That is, the survivor function at time t, specifies the probability that a subject will "survive" for a time t or longer.

An alternative, mathematically equivalent way to specify the probability distribution is through its hazard function, defined as:

$$\begin{aligned} \lambda(t) &= f(t)/S(t) \text{ for } t > 0 \\ &= 0 \text{ for } t \leq 0, \end{aligned}$$

where $f(t)$ is the probability density function corresponding to $S(t)$.

By elementary probability it is possible to show that $\lambda(t)\delta t$ may be thought of as corresponding roughly to the conditional probability that a subject will fail in the next instant of length δt after t, given that failure has not occurred up to time t. This is the sense in which $\lambda(t)$ is a hazard.

There are several reasons for introducing the hazard. The method of maximum likelihood, which is the most widely used procedure by statisticians for devising estimators, depends on what is known as the likelihood function. In survival analysis the hazard is the basic building block used in constructing the likelihood function. Further, the hazard is closely related to the intensity function which is basic to the modern approach to survival analysis, through counting processes. Also, as we shall see in Section 8, the hazard function provides a very convenient means to introduce covariates.

There are two main subdivisions of survival analysis: parametric and nonparametric.

5. Nonparametric Methods

In this section we make no assumptions about the form of either the distribution of the failure times, T_i, or of the censoring times C_i. For instance, we shall not assume that the T_i's follow a Weibull distribution or any other named distribution such as the Normal, which depend on one or more parameters. Hence the terminology "nonparametric". Perhaps the terminology "distribution free" methods would be more appropriate.

We have two main goals in this section:

1. Describe how to estimate the survivor function $S(t)$, and
2. Describe how to compare two survivor functions, $S_1(t)$ and $S_2(t)$ when we have observed data, some of which may be censored, of the type (X_i, δ_i), from each of the two populations with these distributions.

We shall conclude this Chapter with the analysis of a data set, to illustrate the methods of this and other sections of this Chapter.

The Kaplan-Meier (Product Limit) Estimator

Let us suppose that we observe n failure/censoring times, together with their censoring indicators,

$$(X_1, \delta_1), (X_2, \delta_2), \cdots, (X_n, \delta_n).$$

The Kaplan-Meier estimator may be constructed by first focussing on each observed failure time and estimating the hazard at each of these failure times. These estimated hazards are then combined in a natural way.

Let us assume that "true" failures have occurred at:

$$t_1 < t_2 < \cdots < t_k,$$

and suppose we examine what has happened just before a failure time t.

There will be, say r_t subjects, at risk to fail at t. That is, there will be r_t subjects who have survived until time t. Prior to this instant some of the initial cohort (at time $t = 0$) will have failed and some will have been censored; of the n with whom we started, r_t remain. Let d_t denote the number who fail at time t.

The conditional probability of surviving beyond t, is estimated by the proportion of subjects who survive beyond t,

$$\frac{r_t - d_t}{r_t}$$

The next main idea is that to survive beyond u, you must survive up to u and then not fail at u. The probability of this event, $S(u)$, is by conditional probability, equal to:

P(not failing at u given survival up to u) P(survival up to u).

In constructing an estimator of $S(u)$ we start with u and proceed backwards in small steps, at each step estimating the probability of surviving beyond that time point given survival up to that time point.

As an estimate of $S(u)$ we then end up with a product,

$$\frac{r_{t_1}-d_{t_1}}{r_{t_1}} \bullet \frac{r_{t_2}-d_{t_2}}{r_{t_2}} \cdots \frac{r_{t_j}-d_{t_j}}{r_{t_j}}$$

where t_j is the largest observed failure less than u.

The estimator, $\hat{S}(u)$, as we have noted, had its beginnings as early as the 17th century and had been used by actuaries long before 1955 when Kaplan and Meier wrote their important paper in which they systematically explored the statistical properties of $\hat{S}(u)$. For this reason $\hat{S}(u)$ is often termed the Kaplan-Meier estimator though the terminology, product limit estimator, is also used. 6

There are several features of $\hat{S}(u)$:

1. The censoring times play a role in determining the values of r_{t_i}, the number of subjects at risk just before time t_i. In between two observed failure times subjects may be censored thereby decreasing the values of r_t. This is why it is crucial to record both the failure and censoring times AS WELL AS the indicators δ_i.
2. $\hat{S}(u)$ is a step function which is constant in between the observed failure times. It jumps downwards at each failure time. (See the example at the end of the Chapter).
3. It is clear that $\hat{S}(0) = 1$ and that if the last observation is a failure, $\hat{S}(u)$ will be zero for all u beyond this last observation. There is a slight difficulty, though, if the last observation, c, say is censored, for then $\hat{S}(u)$ will have taken its last jump downwards, at the largest observed failure time. This jump, however, would not have carried $\hat{S}(u)$ to zero as there would still remain subjects at risk. The problem is what to do with this "suspended" $\hat{S}(u)$ beyond u. We would badly like to reflect our knowledge that $S(u)$ approaches 0 as u approaches $+\infty$. One way to achieve this property for $\hat{S}(u)$ is to simply define $\hat{S}(u)$ to be zero for all u greater than this last observation, which is censored. This causes $\hat{S}(u)$ to have undesirable statistical properties so that most statisticians prefer to simply leave $\hat{S}(u)$ undefined for u larger than c. This solution is sensible as there is little that can be said about survival beyond c. See Figure 6.2.
4. Having computed the estimated survival distribution using the Kaplan-Meier estimator we would usually summarize the "center" of this distribution by finding its mean and median. Both the estimated mean and median survival time are easy to compute and are routinely provided as part of the output of computer software packages. If the survival

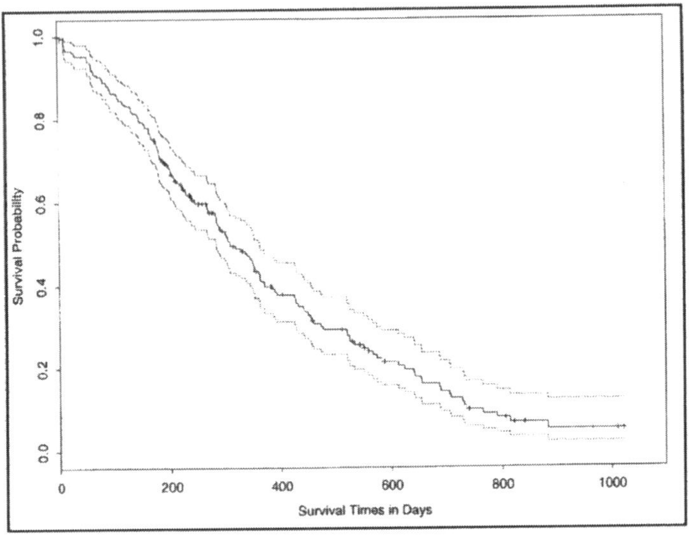

Fig. 6.2. Kaplan-Meier estimator and its 95% two-sided confidence intervals of the survival curve for patients from a lung cancer study.

distribution is suspected to be highly asymmetric it is advisable to give more credence to the estimated median failure time than to the estimated mean failure time, as a measure of centrality.

5. Once $\hat{S}(u)$ has been computed, using a statistical software package, one may estimate the probability of surviving beyond u for any time u. As is usual, however, with the use of point estimators it is advisable to append confidence intervals for the parameters being estimated. This can be done for $S(u)$ for any given value of u, and, of less use, for $S(u)$ for all u simultaneously. We ignore the latter consideration.

The Estimated Standard Deviation of $\hat{S}(u)$ and Confidence Intervals for $S(u)$

Throughout we shall assume that u has been fixed. For example, we may wish to estimate the standard deviation of the estimated probability that a hip prosthesis will last more than 10 years. In short, we may want

$$\widehat{\text{Std. dev}}\left(\hat{S}(10)\right), \text{ or } \widehat{\text{Std. dev}}\left(\hat{S}(u)\right) \text{ for any } u.$$

These standard errors then allow us to find an approximate 100 (1-α)% confidence interval for $S(u)$. One such confidence interval is computed simply by defining:

$$L = \text{lower end point} = \hat{S}(u) - z_{\alpha/2} \text{ Standard deviation}\left(\hat{S}(u)\right), \text{ and}$$

$$R = \text{upper end point} = \hat{S}(u) + z_{\alpha/2} \text{ Standard deviation}\left(\hat{S}(u)\right),$$

where $z_{\alpha/2}$ is the upper $\alpha/2$ point of the Standard Normal distribution.

This confidence interval is symmetric about $\hat{S}(u)$, the Kaplan-Meier estimator at u.

It is clear that for u close to zero, L may be negative while for u close to 1, R may be larger than 1. Since such values for L and R are meaningless it is usual to define $L=0$ in the former situation and $R=1$ in the latter.

An alternative is to construct asymmetric confidence intervals whose end points always lie between 0 and 1. Most software packages provide both symmetric and asymmetric confidence intervals.

6. The Comparison of Two Survival Distributions

Often we carry out a study with the aim of comparing survival between two groups. For instance, we may ask the question, "Is there a difference between the time to failure for two different types of hip prostheses?", or perhaps, more informatively, "What is the magnitude of the difference of the survivor functions at one or more preselected time points?"

For the testing problem, let $S_1(u)$, $S_2(u)$ be the survival functions of the two populations that we wish to compare. Let $\lambda_1(u)$ and $\lambda_2(u)$ be their respective hazard functions. Then the hypotheses that are often tested are:

$$H_0 : S_1(u) = S_2(\mathrm{u}) \text{ for all } u$$

versus

$$H_a : S_1(u) \neq S_2(u) \text{ for at least one } u \qquad (3)$$

The hypotheses (3) are equivalent to:

$$H_0 : \lambda_1(u) = \lambda_2(u) \text{ for all } u$$

versus

$$H_a : \lambda_1(u) \neq \lambda_2(u) \text{ for at least one } u.$$

The data which will have been collected are a sample of possibly censored failure times from each of the two populations:

$$(X_1,\delta_1),\ (X_2,\delta_2),\ \cdots,(X_m,\delta_m),\text{ and}$$
$$(Y_1,\ \delta_1),\ (Y_2,\ \delta_2),\ \cdots,\ (Y_n,\ \delta_n).$$

By far the most commonly used test of the hypotheses (3) is the log rank test or modification of it called the weighted log rank test. The log rank test is also called the Mantel-Haenszel test. The idea is quite simple: the log rank statistic, T, is of the form:

$$T = \frac{\left(\sum_{i=1}^{k} O_i - E_i\right)^2}{\sum_{i=1}^{n} Var(O_i)}$$

where O_i= the number of observed failures in group 1, at the observed failure time t_i, E_i = the expected number of failures in group 1, at observed failure time t_i. This expected number of failures is computed under the hypothesis H_0. That is, under the assumption that the two survival distributions are the same it is possible to calculate E_i, and also:

$$\mathrm{Var}(O_i) = \text{the variance of } O_i.$$

The log rank statistic is therefore based on a comparison between the observed versus the expected number of failures at each observed failure time. Algebraic expressions for E_i and $\mathrm{Var}(O_i)$ are well known but we need not concern ourselves with them here. Suffice it to say that with the information available from our two samples it is easy to compute:

T_{obs} = the observed value of the test statistic T.

This is done routinely in survival analysis software packages.

The test statistic, T has, for large sample sizes m and n, an approximate chi-square distribution with 1 degree of freedom, under the null hypothesis. In short, under H_0, the p-value is defined as:

$$P = P(T > T_{obs})$$

where $T \sim \chi_1^2$.

Notes:

i. If it is suspected that the survivor functions, $S_1(t)$ and $S_2(t)$ cross, the log rank test may have low power. See Figure 6.3. In such a situation even though $S_1(t)$ and $S_2(t)$ are truly different, the logrank test will most likely yield a nonsignificant p-value.
   In the circumstances depicted in Figure 6.3, for some values of i, $O_i - E_i$ will be negative while for others $O_i - E_i$ will be positive. In computing the sum BEFORE we square it, these positive and negative values will tend to roughly balance out, resulting in a small value of T_{obs}.
   The logrank statistic is not of exactly the same form as the familiar chi-square statistic:

$$\sum_{i=1}^{k} \frac{(O_i - E_i)^2}{E_i}$$

   In which the differences are squared before adding. There are statistics, though, of this form that are less commonly used.

ii. If it is suspected that the two survival curves are not "parallel" and cross at some point you may wish to use a weighted logrank test. The idea behind weighted logrank tests is that higher weights, W_i, are assigned to those time points which are of particular interest, either before or after the survival curves cross.
   For example, we may be particularly interested in comparing the survivor functions of two different hip prostheses for times beyond three years. It could be that the survivor curves cross at around three years. See Figure 6.3. Weights that increase as i increases would then be appropriate, as they diminish the effect of the pre 3-year differences O_i-E_i, which are of opposite sign to the post-3-year differences.
   Two of these weighted logrank tests are the Gehan test and the Tarone-Ware test.

iii. There are tests which allow for the simultaneous comparison of more than two populations.

iv. The terminology logrank tests originates from a general class of statistical procedures based on the ranks of observations rather than on the observations themselves. It is possible to show that the tests of the form discussed in

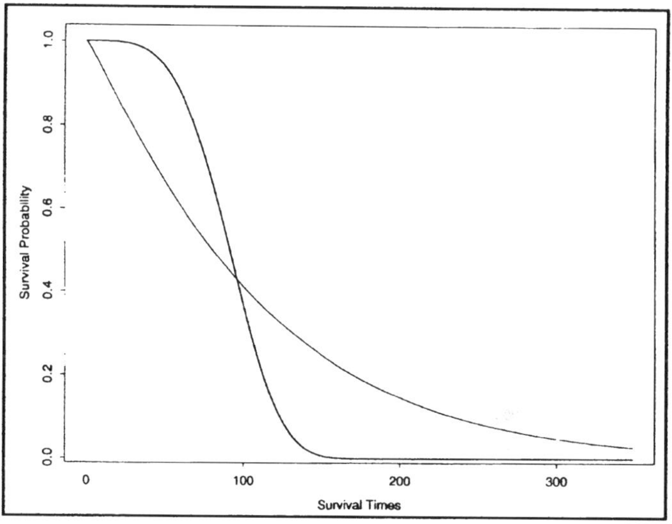

Fig. 6.3. Crossing survival curves.

the section fall into this general class, even though this is not obvious from the form of T.

As was mentioned before, it may be more informative to compute confidence intervals for differences $S_1(t)$-$S_2(t)$ at preselected values of t. In testing, we assume under the null hypothesis, that the survival experience for the two groups is identical (hardly realistic). The confidence intervals allow us to assess the magnitudes of the differences, which are always more clinically relevant than p-values (see Chapter 2), though in analogy to multiple testing, the interpretation of multiple confidence intervals requires some care and a statistician should be consulted.

7. Parametric Models

Sometimes, a researcher who is particularly familiar with the pattern of failures in a certain situation, may wish to use a parametric model for the survivor function. This means that, apart from some unknown parameters, an actual form is imposed on the survivor function or its probability density or its hazard function. For example you may assume that the time to failure of a certain hip prosthesis has a survivor function that is of the form,

$$S(t) = \begin{cases} \exp\left(-\lambda t^{\gamma}\right), & \text{for } t > 0 \\ 1, \text{for } t \leq 0 \end{cases} \tag{6}$$

where λ and γ are unknown positive parameters to be estimated from the data.

This survivor function describes the Weibull distribution which is one of the most important "named" distributions of survival analysis. Other well known distributions in survival analysis are the lognormal distribution, the gamma distribution and the logistic distribution.

The advantage of using a parametric model instead of a nonparametric model is that if your data do in fact come from the parametric distribution that you specify, your inference will be more precise. In the nonparametric setting, for each t, $S(t)$ is unknown and each value of $S(t)$ can therefore be thought of as an unknown parameter. There will, in general, be infinitely $S(t)$'s ("parameters") to be estimated. For a parametric model, for example, the Weibull model defined by (6), once you have estimated the two parameters *lambda* and *gamma*, you can estimate any value of $S(t)$ as:

$$\hat{S}(t) = \exp\left(-\hat{\lambda}t^{\hat{\gamma}}\right), \text{ for } t > 0,$$

where $\hat{\lambda}$ and $\hat{\gamma}$ are the estimated values of λ and γ.

6

The estimated survivor function, $\hat{S}(t)$, will be a smooth curve. (See the end-of-Chapter example.) Confidence intervals about any value $S(t_0)$ will be narrower than their nonparametric counterparts because of the smaller burden of having to estimate fewer parameters.

The disadvantage is that you may not be at all certain which parametric model to use. If you pick the wrong one, say a Weibull, but the data really have arisen from another, say a lognormal, your inference could be seriously flawed. Think of a nonparametric procedure as a Swiss army knife. If you go out into the woods, uncertain what principal task you will encounter, a Swiss army knife (nonparametric procedure) would be useful, as it is versatile. On the other hand, if you knew a priori, that your main task would be to cut branches, you would take a saw (parametric procedure) with you! Yet, if your task turned out to entail using a pair of scissors, a saw would hardly be useful.

We do not describe parametric procedures in any detail in this Chapter, although most software packages allow you to use any of a range of parametric models.

8. Covariates

It is quite possible that you want to examine the effect of one or more covariates on survival or perhaps adjust for the effect of one or more covariates in a clinical trial. For instance, the initial weight of a patient is likely to effect the length of time a hip prosthesis will last. To assess this weight effect you would include weight into your model as a covariate. The same general ideas for covariate specification prevail in survival analysis as for say linear models, with one important difference, the use of time dependent covariates.

In this Chapter we describe briefly the commonest way to build covariates into a survival analysis model and introduce time dependent covariates.

We divide covariates into two main categories, although finer subdivisions are possible.

Fixed Covariates

These are covariates whose values remain constant throughout the period of study. For example, the covariate, gender, may influence survival time of a hip prosthesis, in part because of its correlation with level and type of daily activity, known to be correlated with time to failure of such prostheses.

Time Dependent Covariates

There are covariates whose values change over time. The simplest types are those whose values may be determined at any time in a study as soon as a value at any other time has been specified. Age following surgery is not constant over time (we wish it were!) but changes in a predictable fashion. Incorporating such covariates is straightforward.

Covariates which change randomly as time progresses are harder to include in models. Consider, once again, patients who undergo hip replacement, and whose monthly weights are monitored, since it is believed that these weights influence the time to failure of the prosthesis. The statistical analysis concentrates on the times at which each prosthesis fails. At each such time point we must know not only the weight of the subject whose prosthesis has failed but the weights, at that time, of all those whose prostheses have not yet failed. Clearly, this is a tall order and we usually have to compromise by choosing reasonable proxies for these weights. This particular example illustrates another subtlety. When carrying out a survival analysis with time dependent covariates we are almost always required to summarize the history of these covariates up to time t by a single value. This summarization must make medical sense. The weight of the subject at the time of failure almost certainly does not reflect the effect of that subject's weight on his/her prosthesis since surgery. It is likely, rather, that the changing weight pattern effects the time to failure of the prosthesis. The researcher and statistician must replace this complete weight history with, say, the average of the monthly weights up to that time or perhaps the maximum of the weights since surgery.

When time dependent covariates are called for it is definitely time to consult a statistician for advice.

How Are Covariates Included?

We begin with the imprecise notion that one or more covariates may affect the survival times under study. In order for this qualitative idea to be useful it must be formulated so that we can say exactly how survival is affected by covariates. Three possibilities immediately come to mind:

1. The failure times themselves are directly influenced by the covariates. One common assumption is that the natural logarithm of the failure time, T, log T, can be expressed in a familiar regression form:

   $$\log T = \alpha + \beta \times \text{covariates} + \text{"random error"}.$$

   Here, though, the random errors are not Normally distributed. Put another way, this model assumes that the effect of covariates on the failure times (as opposed to the logarithms of the failure times) is to multiply that is, to accelerate or decelerate the failure times. Hence, these models are called accelerated failure time (AFT) models.
2. The covariates affect the failure times indirectly by being built into the hazard function. As we shall see shortly, this can be done in an intuitively reasonable manner and, at the same time a rich statistical methodology has developed from such models.

   It is interesting to note that for certain choices of the distribution of the random error in (1) and the hazard function in (2) the direct and indirect approach are equivalent. In fact, some software packages present the output for certain parametric models in AFT form, so that this should be borne in mind when interpreting the parameters.

3. The covariates affect the failure times indirectly by being built into the survivor function. It is difficult to do this in a way that leads to easy interpretation of the covariate effects, unless one refers to the hazard function. The more natural starting points are, therefore, the approaches described under (1) or (2).

The Cox Proportional Hazards Model

We sketch the main features of the approach (2), emphasizing a particularly flexible and popular model due to Cox. The idea is simple. Let $\lambda(t;z)$ be the hazard of failure at time t, of an individual with an observed single covariate z. Let there be some "baseline" hazard, $\lambda_0(t)$, common to all subjects in the study. In a proportional
6 hazards model we suppose that $\lambda(t;z)$ can be expressed in the form:

$$\lambda(t;z) = \lambda_0(t) \times f(z),$$

where $f(z)$ is some function of the covariate z.

That is, the hazard at time t, of a subject with covariate z, is proportional to the baseline hazard $\lambda_0(t)$. The function of the covariate, $f(z)$, is the constant of proportionality as it does not depend on t.

Now we could specify a parametric form for $\lambda_0(t)$ (for example, a Weibull hazard) or we could leave its form unspecified to allow greater flexibility. We shall take the latter approach. Next, we can specify a form for $f(z)$. That is we describe how we think the covariate z enters the picture. The most commonly used form for $f(z)$ is

$$f(z) = \exp(\beta z).$$

Our model then becomes

$$\lambda(t;z) = \lambda_0(t)\exp(\beta z), \qquad (5)$$

and is known as Cox's proportional hazards model. This is an example of a semiparametric model as $\lambda_0(t)$ is not given a parametric form, while $\exp(\beta z)$ specifies that the covariates enter in a linear fashion in the exponent, through βz, with unknown parameter β.

Several features of the proportional hazards model are noteworthy.

1. There is no reason to restrict attention to one covariate at a time. If we are interested in the simultaneous effect of the covariates $z_1, z_2, \cdots, z_k$, such as weight, gender and type of prosthesis, then the model becomes:

$$\lambda(t; \mathbf{z}) = \lambda_0(t)\exp(\beta_1 z_1 + \beta_2 z_2 + \cdots + \beta_k z_k).$$

2. We are now ready to interpret the parameters β_i. Imagine two subjects whose covariates other than the ith, are identical. Suppose that their ith covariates differ by one unit. Remembering that $\lambda(t;z)$ represents the risk (hazard) of failing at time t, e^{β_i} is then the relative risk of failure for these two hypothetical subjects. It is in this sense that β_i describes the effect of covariate z_i. In other words, β_i describes the change in the log of the relative risk per unit change in the covariate z_i, keeping the other covariates

fixed. The analogy with, and, at the same time, the difference between the interpretation of a regression coefficient, β_i, in a linear regression model is obvious: there, β_i is interpreted as the change in the expected value of the dependent variable per unit change in the covariate z_i, keeping all other covariates fixed.

3. Because $\lambda_0(t)$ has no parametric form, its infinitely many possible values are not easily dealt with by standard statistical procedures; our focus is on the βs, the covariate effects and $\lambda_0(t)$ gets in the way. Cox cleverly devised his "partial likelihood function" to overcome this difficulty, and, remarkably, many of the results of standard likelihood theory continue to hold when partial likelihoods are used.
4. Although the main purpose of using Cox's proportional hazards model is to examine the effects of covariates on survival, it is possible to estimate the survivor function at any time point for given covariate values. Therefore, a future subject's survival probabilities may be estimated once his/her covariates become known. Alternatively, one may estimate the survivor function at covariate values representative of some group, if such values make sense.
5. The sequence of model choice, parameter estimation and inference, as well as model fit assessment through residual examination, pertain in survival analysis as in a regression analysis. At the same time, the identical warning flags fly, as a reminder about the pitfalls of the blind use of automated regression procedures such as forwards or backwards regression. Only the researcher can decide which is the most appropriate subset of variables to use in a model. The set of competing models should be small and arrived at after careful consideration of the underlying science and not through a succession of p-values.

9. An Example of a Survival Analysis

In this study we apply the methods that we have outlined in this Chapter. The analyses presented have been done with a view to illustrating the various topics of this Chapter.

We abstracted data from the records of patients with malignant brain tumors who had undergone initial surgery at the Montreal Neurological Hospital, between 31 January, 1989 and 22 October, 1996. The general aim was to "say something" about how long such patients survive, to compare the survival times of well defined populations of patients with malignant brain tumors and to examine the effects of several covariates on survival (Table 6.1).

Most studies would begin with a general goal, but before collecting the data the aims should be sharply defined so that the data collected will, in the end, be useful. Also, it is not correct statistical practice to search the data for hypotheses to test, because they are suggested by the data. We were specifically interested in:

1. Patients who, at the time of initial surgery, were diagnosed as having either Grade 3 or Grade 4 astrocytoma (that is anaplastic astrocytoma or glioblastoma multiform, respectively);

Table 6.1. The summary statistics of the data

Age at surgery in weeks	min: 878; max: 4026; median: 2708; mean: 2703
Number of females	34+ see note below
Number of males	27+ see note below
Number of Grade 3	18
Number of Grade 4	47
Total number of patients	65 Note: for 4 patients gender was not abstracted

6

2. estimating the survival distribution of the (abstract) population of patients who would have been treated at the Montreal Neurological Hospital during the study period;
3. nonparametrically comparing the survival distributions of males and females as well as patients with Grade 3 versus Grade 4 astrocytomas, and in
4. assessing the simultaneous effects of gender and tumor grade on survival.

With these modest goals in mind, we prepared a simple data abstraction form on which were recorded

1. the date of birth,
2. the date of surgery,
3. the gender of the patient,
4. the grade of tumor at the surgery, and
5. the date of death or, if the patient was lost to follow-up or was alive at the end of the study, the date last known to be alive.

Following the end of the accrual period 22 October, 1996, all patients included at that time, were retrospectively followed until March 10, 1998, when the study ended.

For the sake of simplicity other features of the data, which would certainly be examined in a more complete analysis, have not been displayed.

There were two sources of censoring in this study:

1. patients were lost to follow-up and it is assumed that their loss to follow-up was not related to their subsequent survival, and
2. patients still alive at the end of the study were considered censored at that time. In the former case, clearly the censoring times are independent of the survival times. In the latter case this is true as well, if we assume, reasonably, that patients enter the study (have initial surgery) at times independent of their disease severity. To understand this point, suppose that patients with more severe disease tended to enter the study towards the end of the accrual period. Then their censoring times (at the end of the study) would be shorter and dependent on their shorter survival times.

One final point before the analysis. If the patients enter the study at times independent of one another, again, a reasonable assumption, the between patient survival/censoring times are independent.

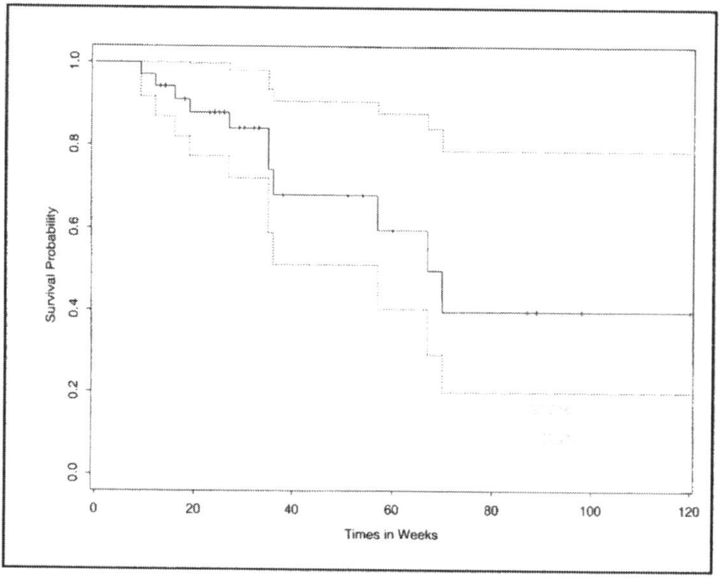

Fig. 6.4. Kaplan-Meier estimator and its 95% two-sided confidence intervals of the survival curve for astrocytoma patients.

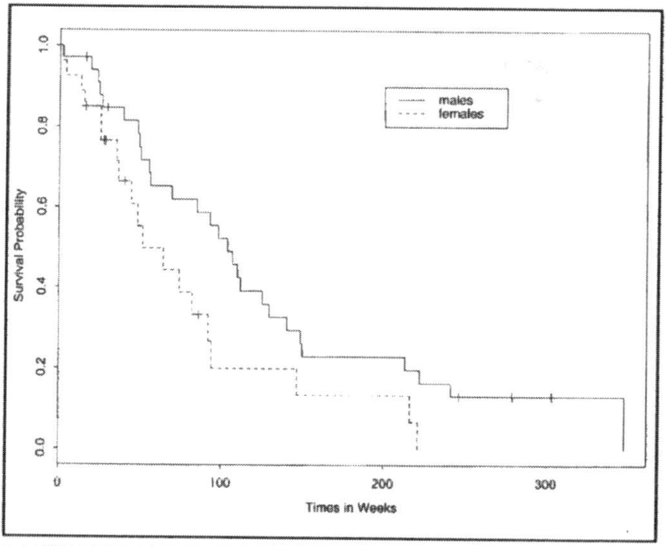

Fig. 6.5. Kaplan-Meier estimators of the survival curve for astrocytoma patients by gender.

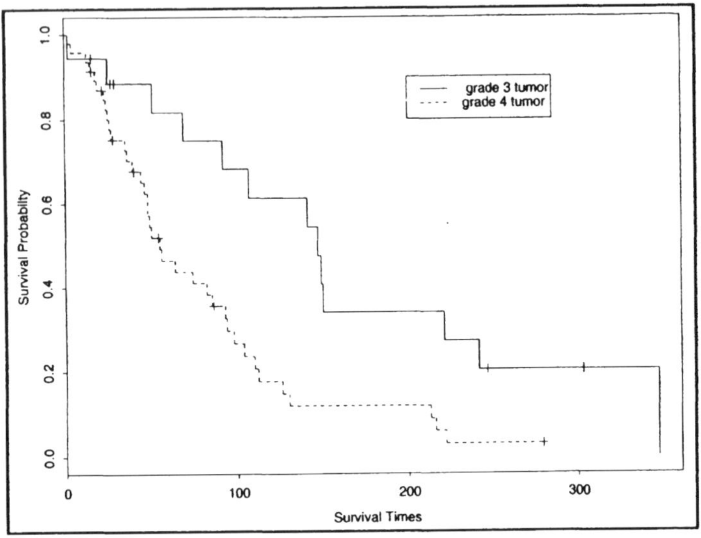

6

Fig. 6.6. Kaplan-Meier estimators of the survival curve for astrocytoma patients by tumor type.

The Analysis

Kaplan-Meier survivor functions were computed for

1. all patients combined, see Fig. 6.4,
2. males and females separately, see Fig. 6.5,
3. Grade 3 and Grade 4 tumors separately, see Fig. 6.6, and
4. the four possible combinations of gender and tumor grade separately as it was believed that there might be an interaction between tumor grade and gender. This means that, although we would expect a difference in survival between patients with Grade 3 and Grade 4 tumors, it was also supposed that this difference (that is, grade effect) would not be the same for males and females. This hypothesis is suggested by Figure 6.8, in which the "male survivor" curves are further apart than their female counterparts. In practice, any hypotheses should be determined independently of the current data, so that Figure 6.8 should not be used as a justification for including this particular interaction. We have done so for illustrative purposes only. See Figure 6.7.

In Figure 6.4 the darker line is the estimated survival function. The outer lines form the 95% confidence limits for the true survival distribution. Even allowing for the "correct" interpretation of confidence intervals, the most we can say is that at each time point u^* on the horizontal (time) axis, the points above and below $\hat{S}(u^*)$ are 95% confidence limits for $S(u^*)$. The confidence statement does not apply simultaneously for all u. Table 6.2 provides the failure times (in weeks) following

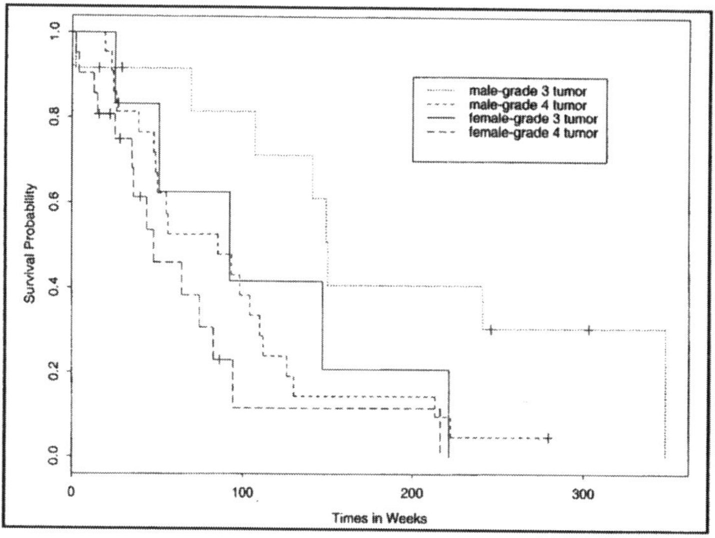

Fig. 6.7. Kaplan-Meier estimators of the survival curve for astrocytoma patients by gender and tumor type combinations.

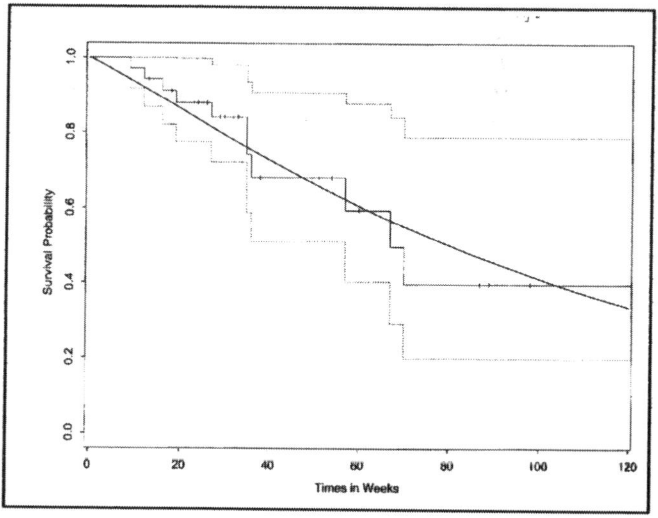

Fig. 6.8. Weibull curve overlaid with Kaplan-Meier curve.

surgery, the size of the risk set at these times, an indictor of 1 (died) or 2 (censored), the estimated survival probabilities at these times, the estimated standard deviations of $\hat{S}(u)$ for each failure time u, and the upper and lower confidence limits.

Figure 6.5 seems to point to a possible difference in the survival of males and females, with females having shorter survival.

Figure 6.6, not surprisingly, indicates that the outcome for those with Grade 4 tumors is worse than for those with Grade 3 tumors.

From both Figures 6.5 and 6.6 we see that the estimated survival curves are roughly "parallel" to one another, which means that an unweighted logrank test is appropriate to use in both instances.

From Figure 6.7 we see that the difference is survival between Grade 3 and Grade 4 tumors is larger for males than for females, indicating a possible interaction between gender and grade of tumor.

6

Figure 6.8 displays the estimated survival distribution for all patients combined using the Weibull model. In contrast with the nonparametric Kaplan-Meier estimator, the curve is smooth and fits reasonably closely to the Kaplan-Meier estimator.

Unweighted logrank tests were carried out to test two sets of hypotheses:

$$\begin{aligned} &H_0 : S_M(u) = S_F(u) \text{ for all } u \\ vs. \quad &H_a : S_M(u) \neq S_F(u) \text{ for at least one } u, \end{aligned}$$

where $S_M(u)$ and $S_F(u)$ are the male and female survivor functions respectively, and

$$\begin{aligned} &H_0 : S_3(u) = S_4(u) \text{ for all } u \\ vs. \quad &H_a : S_3(u) \neq S_4(u) \text{ for at least one } u. \end{aligned}$$

Both tests were significant ($p = 0.0347$ and $p = 0.00447$ respectively), consistent with our observations about the estimated survival distributions.

Although most standard survival analyses in which two strata are compared nonparametrically, are based on a form of the logrank test, we feel that it is, perhaps, more informative to present confidence intervals for the differences in survival at a few preselected time points. We have no objection to the logrank test per se, only to the principle of testing the point hypothesis that two survival distributions are identical, and reporting a p-value. Recall from Chapter 2 a p-value cannot be directly interpreted as a measure of belief in this hypothesis. Instead, by deciding which "milestones" in survival are important we may by computing confidence intervals, obtain an idea of the magnitude of the differences as well as an indication if the differences contain the point 0. If such a confidence interval includes 0, we would suspect that there may be no difference in survival between the two groups at that time point.

In Table 6.2 we therefore present 95% confidence intervals for the differences $S_M(u)$-$S_F(u)$, and $S_3(u)$-$S_4(u)$ for $u = 25$, 50 and 150 weeks respectively. The results are presented below.

It was decided a priori to compare two models using Cox's proportional hazards model.

Model 1 includes the covariates sex, grade of tumor and an interaction term between these two covariates.

Model 1:

$$\lambda(t) = \lambda_0(t)\exp(\beta_1 x_1 + \beta_2 x_2 + \beta_3 x_1 x_2)$$

where

$$\begin{aligned} x_1 &= 0 \text{ if male} \\ &= 1 \text{ if female, and} \\ x_2 &= 0 \text{ if Grade 3} \\ &= 1 \text{ if Grade 4} \end{aligned}$$

and the term "x_1x_2" models the interaction between the covariates gender and tumor grade.

Model 2 excludes the interaction term.

Model 2:

$$\lambda(t) = \lambda_0(t)\exp(\beta_1 x_1 + \beta_2 x_2).$$

Likelihood ratio test =11.8 on 2 df, p=0.003.

From Table 6.3 it can be seen that the likelihood ratio test is highly significant. This means that there was a significant difference between the estimated model with just the baseline hazard and one including all three terms of Model 1. On the other hand, only one of the estimated parameters (tumor grade), is significantly different from zero. There is no real contradiction, as the likelihood ratio test compares the effects of covariates taken as a group. The estimated parameters of Model 2 are, therefore, difficult to interpret.

6

When the interaction term is omitted, the likelihood ratio test is still significant and the estimated gender and tumor grade effects are now significant. It appears as if both gender and grade of tumor are predictors of survival.

Closing Remark

The model fitting using Cox's proportional hazards model could have been carried out using a fully parametric model such as a Weibull model adapted to accommodate covariates. It is recommended that one begin with a simple nonparametric approach, proceed to a parametric approach if covariates are to be accounted for and assess the adequacy of this first "stab". If this appears inadequate, proceed to Cox's

Table 6.2 Confidence Intervals for $S_M(u)$-$S_F(u)$ and $S_3(u)$-$S_4(u)$

u	$S_M(u)$-$S_F(u)$	Lower 95% Confidence Limit	Upper 95% Confidence Limit
25	0.082	0.044	0.120
50	0.220	0.170	0.270
150	0.162	0.127	0.197
u	$S_3(u)$-$S_4(u)$	Lower 95% Confidence Limit	Upper 95% Confidence Limit
25	0.085	0.046	0.124
50	0.297	0.247	0.347
150	0.252	0.194	0.310

Tables 6.3 and 6.4 show the results of fitting these two models, respectively.

	coef	exp(coef)	se(coef)	z	p
gender	0.916	2.498	0.597	1.534	0.130
tumor	1.061	2.888	0.451	2.353	0.019
interaction	-.0367	0.693	0.688	-0.534	0.590

Likelihood ratio test =12.1 on 3 df, p = 0.007.

	coef	exp(coef)	se(coef)	z	p
gender	0.642	1.9	0.314	2.05	0.04
tumor	0.915	2.5	0.349	2.62	0.00

6

proportional hazards model keeping in mind that even this flexible model may not be appropriate.

Acknowledgement

We thank Dr. Richard LeBlanc of the Montreal Neurological Institute for kindly providing the data for the example. Ms. Irene Leboeuf expertly abstracted the data from patient files. She impressed upon us the importance of realizing that each set of data represented a person facing a serious and, in most cases, fatal disease, with remarkable courage and dignity.

Selected Readings

1. Boddie AW, Jr, Warso M; Briele H et al. Multimodal-therapy breast salvage in the urban poor with locally advanced cancer. Arch Surg Vol 1996; 131:424-428.
2. Collet D. Modelling survival data in medical research. Chapman and Hall. Comment: For those with some statistics training. 1994
3. Halley E.. An estimate of the degrees of mortality of mankind, drawn from curious tables of the births and funerals at the City of Breslaw; with an attempt to ascertain the price of annuities upon lives. Philos Trans Roy Soc London, 1693; Vol. 17:596-610.
4. Johnson MW, Washburn K, Freeman RB et al. Hepatitis C viral infection in liver transplantation. Arch Surg 1996; 131:284-291.
5. Kaplan EL, Meier P. Nonparametric estimation from incomplete observations. J Amer Stat Assoc 1958; 53:457-481.
6. Harris EK, Albert A. Survivorship analysis for clinical studies. Marcel Dekker Inc. 1991. Comment: At a somewhat lower level than the book by Collet, but still requiring some familiarity with statistical methods.
7. Wanebo HJ, Chu QD, Vezeridis P et al. Patient selection for hepatic resection of colorectal metastases. Arch Surg 1996; 131:322-329.

CHAPTER 7

Brief Introduction to Meta-Analysis

Lawrence Joseph

1. Introduction—Why Consider Meta-Analysis?

According to Last's A Dictionary of Epidemiology, meta-analysis is:

> *The process of using statistical methods to combine the results of different studies. The systematic, organized and structured evaluation of a problem of interest, using information from a number of independent studies of the problem.*[1]

Thus meta-analysis is helpful in synthesizing the information about a particular edical issue via statistical analysis, especially when more than one study has been rried out that relates to the question of interest. Since one should ideally consider of the available information about any issue before making any clinical decisions, d since meta-analysis is able to summarize information from several data sets, it uld seem that meta-analysis should be considered as a cornerstone of medical earch and medical decision making. In some ways this is true, but unfortunately, many problems that are associated with meta-analysis make its use controversial best, and a poorly conducted meta-analysis can produce results that are down-ht misleading. Nevertheless, the use of meta-analysis is on the rise, and a carefully rformed and reported meta-analysis can produce useful results. In addition, new tistical methodologies are continually being developed, many of which address ecific problems that have been raised about meta-analysis.

In this Chapter, we will provide an overview of the various methods that have en used in meta-analysis, including some of the most common statistical models t have appeared in the literature. We will also discuss several of the problems and ses that are difficult to avoid in carrying out a meta-analysis, and indicate how y can be minimized. In particular, Section 2 discusses the specification of the estion that the meta-analysis will focus on, and the subsequent selection of stud- whose data will be included in the analysis. Simple fixed effects statistical tech-ues for carrying out a meta-analysis will be covered in Section 3, and more complex dom effects techniques will be discussed in Section 4. Throughout, we will apply of the techniques discussed in Sections 3 and 4 to a research question first consid-d by Pocock et al.[2] These authors summarized the results from a collection of dies comparing coronary angioplasty (PTCA) to bypass surgery (CABG) for ients with severe angina. This example also provides us with an opportunity to phasize the care in which meta-analytic results must be interpreted, especially en deciding upon the context to which the results can be applied. This example is

gical Arithmetic: Statistics in Surgery, by Lawrence Rosenberg, Lawrence Joseph, Alan Barkun. ©2000 Landes Bioscience.

summarized in Section 5, and Section 6 concludes with a summary and suggestions for further reading.

As this Chapter presents a variety of statistical techniques for meta-analysis, a solid grounding in the basics of statistical analysis is an essential prerequisite. Therefore, we suggest reading through Chapter 2 before tackling the material in this Chapter.

2. What Is the Question, and Which Studies Should Be Included?

Two of the first steps in carrying out a meta-analysis are the specification of the research question that is to be investigated, and the choice of which studies to include from among all those which address the research question of interest. Neither of these tasks are as straightforward as one might first assume. Below we discuss each of these in turn.

2.1. Specifying the Research Question

In choosing a research topic, many factors must be considered in order to make the question as precise as possible. For example, suppose one wishes to investigate any outcome differences between PTCA and CABG in patients with severe angina. First, the definition of "severe angina" must be made precise. As for outcomes, should one consider cardiac deaths only or also include nonfatal myocardial infarction? What about the rates of subsequent revascularizations? Including composite outcomes, such as cardiac deaths or nonfatal myocardial infarction have the advantage of providing more events, and hence often provide increased statistical accuracy in estimating differences between treatments. This increase in statistical accuracy, however, comes at the steep price. If one finds a difference in a combined outcome, one cannot be sure if both or only one of the outcome rates differ between treatments, and this distinction may be important for clinical decision making. For example, a cardiac death is a more serious outcome than a nonfatal myocardial infarction. If cardiac deaths in fact occur at the same rate in both treatment groups but nonfatal MI rates differ, this may be less serious than if cardiac death rates differ. Conversely, if no differences are found between treatment arms for a combined outcome, one cannot be sure whether differences in both outcomes, but in opposite directions, have cancelled each other out. Therefore, if one will use a combined outcome, it is important to also consider each outcome alone in separate sub-analyses.

In a related issue, should one compare the rates in the two groups in terms of the difference in the absolute event rates, or use relative risks or odds ratios? Absolute differences are often more clinically relevant, but if studies have different follow-up times, or if case-control studies are included, for example, only relative risks or odds ratios may be valid or easily available. What age range should be considered? This may be important if the success rates or adverse event rates of the different procedures can change with age. Complicating this issue further is that it is rare for two studies to consider exactly the same age range, so clinical judgement is required to decide if the age differences are such that combining results is reasonable. Other patient characteristics such as aspects of their medical histories might be of importance, as they may also differ markedly from study to study. Thus there are issues of both validity and generalizability that must be considered in narrowing down which populations to include. What about follow-up time? Should one consider events in the next year? Two years? What if there are differences in the ways that the procedures are performed from study to study? Does one need to specify the procedure, or do all of them produce similar event rates, and so can be combined in a straightforward manner? In drug studies, one must consider the dosage or the range of dosages that is acceptable.

In summary, in phrasing the research question, many issues about the procedures and populations to which the procedures are applied need to be carefully considered. Including too narrow a range of procedures or population characteristics hampers generalizability, and may exclude many important studies. On the other hand, inclusion criteria that are too wide may lead to the combining of studies with heterogeneous effects, which may make the meta-analysis results more difficult to interpret. A finding in one direction or another may not apply to all subgroups, and conversely, an overall lack of a positive effect does not rule out that the intervention may still be usefully applied to some subgroups. The difficulty is that the numbers of patients in many or even all of the subgroups of interest will be small, precluding accurate estimation of differences in outcomes within these subgroups. Careful thought about these issues is therefore essential to a well planned meta-analysis.

2.2. *Which Studies Should Be Included?*

Once one specifies the research question, the next task is to find all studies that are related to this question, and choose from among them those that will be included in the meta-analysis. We next consider these two questions.

2.2.1. Finding all Potentially Relevant Studies

Most hunts for relevant studies will start with a formal keyword and topic search in one or more databases of the medical literature. Of course, putting in a key word like "myocardial infarction" will bring up literally thousands of articles, so that search strategies usually need to be more refined. For example, one might begin with the keywords of coronary angioplasty, coronary bypass surgery, and angina. If we were considering only randomized clinical trials, we might add that term as an additional keyword.

Once the relevant articles are gathered, the references from each of these articles can be perused for any missing items of interest. For example, there may be review articles or even previous meta-analyses. Other sources of information can include government agencies, conference presentations, or conversations with experts in the field, who, for example, may be aware of ongoing or unpublished studies. It is important to find all relevant unpublished studies in order to avoid publication bias. This bias can occur if several studies of the issue of interest have been carried out, but only those with "positive" findings are submitted or accepted for publication. As a simple (and perhaps somewhat unrealistic) example, suppose that there are truly no differences between the outcomes of patients with a certain medical condition given two drugs. If 20 trials of these two drugs are carried out, by chance alone, we would expect that one of these 20 trials would be "statistically significant" at the 5% significance level (see Chapter 2). Surely it would give a completely false impression if this "positive" trial were to be the only one published! While this example is somewhat exaggerated, there has certainly been a tendency for researchers to prefer submitting articles with "positive" findings, and a tendency for most medical journal editors to accept such articles in preference to "negative" studies. This so-called "file drawer problem" is often raised as an objection to the validity of meta-analyses.

To summarize, the goal of the search for previous relevant research is not necessarily to find all articles ever written on the topic of interest, but rather to find all important sources of data that address the question of interest.

2.2.2. Which Studies to Include?

Once all potentially relevant articles have been gathered, the next task is to decide which will be formally included in the meta-analysis. First of all, only studies with original data (or corrections to these data!) which directly impact on the question of interest need to be included. Should one only include randomized trials, or also consider observational studies? Randomized trials may be freer from bias, but in some cases, there may be no randomized trial data available. This could occur, for example, when it is unethical to randomize patients, such as in a study of the effects of smoking or exposure to second hand smoke. Because of the specialized protocols and inclusion/exclusion criteria, randomized trials are often more difficult to generalize to all potential patients. Even more difficult is to judge the quality of each study from the article or other source where it is reported. Mixing poor studies with better conducted studies can create biases and imprecisions that can serve to either dilute or exaggerate the sizes of effects of interventions. We have already discussed publication bias, which can increase the apparent effect of an intervention. Conversely, including a study where an intervention has not been correctly implemented can dilute the true effect of the intervention when it is properly administered.

By looking closely at the features of each study, one can determine what sources of potential biases may be present (see Chapter 1), and what degree of confidence one can place in the results. In investigating bias, both internal bias (for example, noncomparability between patients in the two treatment arms in a study) and external bias (for example, strict inclusion/exclusion criteria that may limit generalizability) are of importance. One also needs to consider the degree to which the study design and main study questions of each article match those of the primary research question. What impact, for example, is expected if a drug dosage is slightly different from that which has become standard today? What if the patient population in the study is more or less severely diseased than those to whom we wish to apply the results of the meta-analysis? These are not easy questions to answer, especially if there is a paucity of results that do exactly match the desired research question. The question of whether a given study design is "close enough" to what would be considered ideal is central to the debate about the usefulness of meta-analysis. If small deviations in protocols can potentially change the results in a clinically meaningful way, one must proceed with a meta-analysis with great caution. This is an especially difficult point, since the degrees to which results can be changed by differences in designs are rarely clear.

Aside from the above concerns about bias, there are design specific issues that affect study quality. One of the most obvious design issues is sample size. As discussed in Chapter 2, all other things being equal, larger studies provide more precise estimates of effect sizes than smaller studies. Aside from size, many other design features, including the choice of randomization technique, accuracy of measures and in determination of outcomes, blinding, and length of follow-up (see Chapter 4 for a discussion of the issues pertaining to clinical trials) can all have an impact.

There is a large literature on bias and quality of studies in meta-analysis. A full discussion of this topic is beyond the scope of this Chapter. The interested reader is referred to the books mentioned at the end of this Chapter and the references therein for further details.

3. Fixed Effects Meta-Analytic Methods

Once one has specified the research question, found all relevant articles, judged each article for quality, and made a decision about which articles to include in the meta-analysis, the next step is to analyze the data. Recall that the major goal of this analysis is to combine the results from the different studies so that overall conclusions can be drawn. This section will consider simple methods of meta-analyses, while the following section will consider more complex models.

Before discussing the analysis techniques, we will first present the data from the previous meta-analysis concerning two different revascularization procedures for patients with severe angina.[2] Table 7.1 presents a summary of the results from this article. These authors considered 8 randomized clinical trials comparing coronary angioplasty (PTCA) with bypass surgery (CABG), including a total of 3371 patients with an average follow-up time of 2.7 years. While here we will focus on the combined outcome of cardiac death and nonfatal myocardial infarction, these authors also looked at rates of other cardiac interventions following the initial treatment, and the prevalence of angina in subsequent years. In addition, they performed separate analyses for multivessel and single vessel disease patients. Of course, all of these outcomes and patient groups should be considered when comparing the interventions, although the data in Table 7.1 will suffice for our illustrative purposes.

Looking at the results in Table 7.1, several preliminary observations can be immediately made:

1. There is a range of sample sizes among the 8 clinical trials, from relatively small trials (127 subjects) to relatively large trials (1054 subjects). This implies that the precision of the relative risk (RR) estimates also vary, as reflected in the widely differing lengths of the confidence intervals.
2. The mean follow-up times vary greatly, from 1 to almost 5 years. Thus the proportions of events in each trial can also expect to vary, as, all else being equal, longer trials can be expected to have more events. Nevertheless, the trial with the highest rates (up to 17% in the CABG group), King et al,[5] is almost two years shorter than the longest trial, the RITA trial,[4] which has a much lower rate (about 6 or 7%). Thus follow-up time alone does not fully explain the observed rate variations in the trials; one or more factors such as different trial settings, populations, and variations in procedures, as well as random variation play a role in determining these rates. One must return to the original individual study details and consider all of these factors when drawing conclusions from this meta-analysis.
3. The point estimates of the relative risks vary from study to study, in both magnitude and direction. Thus it is difficult to tell if the overall results indicate a RR above or below 1. Furthermore, all confidence intervals include the value of RR = 1.

In carrying out a meta-analysis, as in all statistical analyses, various assumptions must be made. One of the major assumptions in a meta-analysis concerns the parameter of main interest, which in our example is the relative risk. If we believe that the relative risks from all studies are in fact the same, then a simpler meta-analysis technique can be used compared to if we believe that the relative risks vary from study to study. Note that it is the true (unobserved) value for the RR, not the calculated point estimates, that must be the same. Thus one cannot necessarily decide that the parameters differ between studies from looking at the point estimates alone, since the observed values are subject to random fluctuations about their true values. No two study settings are ever exactly the same, because populations, inclusion and exclusion criteria, specifics of the interventions, and so on, always differ from study to study. Therefore, assuming that the relative risks from two or more studies are exactly the same seems unreasonable. Nevertheless, if they can be expected to be very close, either because the study settings did not differ greatly across studies or because one expects the effects not to greatly depend on situation specific variables, then this "fixed effects" model may be a reasonable approximation to reality. In most other cases, where effects can be expected to differ at least a bit from study to study, a so-called "random effects" model is preferable, since such a model explicitly accounts for between study differences. These more complex models will be discussed in Section 4, while this section discusses three different fixed effects models. From a frequentist viewpoint, we will discuss the pooled and the variance weighted fixed effect approaches. We will also look at a fixed effects Bayesian approach. See Chapter 2 for a discussion of the philosophical and practical differences between the frequentist and Bayesian approaches to statistical analyses.

3.1. Meta-Analysis Using a Pooled Data Approach

The pooled data approach is the most simple (some might say simplistic!) of all meta-analytic approaches. To apply this method one acts as if all of the data comes from a single large experiment, and analyses the data using the appropriate "single study" technique. In the meta-analyses of Pocock et al,[2] one can form a numerator by summing the numbers of events across each study in each of the PTCA and CABG treatment arms, and similarly sum across studies to obtain the total numbers of patients in each study arm. From Table 7.1, these totals would be 127 cardiac death or nonfatal MI events in 1661 patients in the CABG group, and 136 events in 1710 patients in the PTCA group. We can then calculate a relative risk and corresponding confidence interval using the methods for single studies, as described in Chapter 2, Section 5. Thus the estimated pooled relative risk would be:

$$\hat{RR}\frac{\frac{136}{1710}}{\frac{127}{1661}} = 1.04$$

with a confidence interval of (0.83, 1.32).

Pooled meta-analytic estimates can similarly be calculated for other measures, such as risk differences or mean differences. Use of this method implies that the rates (or means) are the same within each treatment arm of each study. Clearly this is not a reasonable assumption here, since even if the rates would theoretically be the same, the length of follow-up of each study varied, so that the reported rates cannot possibly be the same. Therefore, this method is inappropriate for this study, and should at least be seriously questioned in most other situations.

3.2. Meta-Analysis Using a Variance Weighted Approach

All meta-analysis techniques are, at least in a loose sense, a weighted average of the results from the individual studies. The pooled analysis of the previous section, for example, can be considered to be a weighted average where the weights are obtained directly from the sample sizes from each study. If all studies provided equally reliable estimates and all were equally free from bias, then a simple mean (or average) of the results would seem to be appropriate as an overall estimate of the effect of the intervention on the outcome of interest. It is usually the case, however, that some studies provide more reliable estimates than others. This can happen, for example, if some studies have larger sample sizes than others, and thus provide estimates with smaller standard errors (again, see Chapter 2 for definitions of these basic statistical terms). Here a simple average is sub-optimal as an overall estimate of the effect, since we would somehow like those studies with more precise estimates to "count" for more than those providing less precise estimates. In other words, a weighted average giving more weight to more precise studies would be preferred to a simple average. As the name implies, in a variance weighted approach, the weights we use in the weighted average are related to the variance of the estimates from each particular study. The smaller the variance, the more precise an estimate one obtains from an individual study, so the more weight it should receive in calculating the overall estimate of the effect of interest.

If we let the weight $w_i = \frac{1}{var_i}$ be the reciprocal of the variance from study i (so small variances imply large weight, and vice versa), then the variance weighted meta-analytic estimate, $\hat{E}$ of the overall effect E of interest is given by:

$$\hat{E} = \frac{\sum_{i=1}^{k} w_i \times \hat{E}_i}{\sum_{i=1}^{k} w_i} \quad (1)$$

Here k is the total number of studies included in the meta-analysis, and study i has estimated effect $\hat{E}_i$.

Note that E can be any effect of interest, such as a relative risk, an odds ratio, a between treatment difference in means or in rates, and so on. In our case we are looking at the relative risk, so that outcome E is the RR, and the weights are related to the estimated variance of the RR (see Chapter 2, Section 5). Formulae for the variance of the overall estimate ($\hat{E}$) are also available, see, for example, Chapter 19 of Cooper and Hedges.[3]

Applying equation (1) to the data from Table 7.1 gives the results in the second line of Table 7.2. The variance weighted point estimate of the relative risk is 0.88. This means that there is a 12% decrease in the rate of cardiac death or nonfatal myocardial infarction following CABG versus PTCA, but note that the 95% confidence interval (0.42, 1.34) includes the null value of 1, and does not rule out an effect in the direction opposite to the point estimate. Since a risk reduction of

58% (corresponding to the lower limit of the confidence interval of 0.42) would presumably be of great clinical interest if it were the true value, one concludes that this analysis is inconclusive (see Figure 2.8 of Chapter 2), since the null value is included in the confidence interval, but a clinically important effect cannot be ruled out.

Like the pooled method, implicit in variance weighting is the assumption that the effects are identical across studies, so that there are no internal or external biases in any of the studies that might make their effects different. In other words, in variance weighting, one still retains the usual fixed effects assumption of equal relative risks across studies, an assumption that may or may not be reasonable. While in theory one can devise a test for homogeneity of relative risks across studies, these tests are subject to all of the problems associated with *p*-values (see Section 4 of Chapter 2), especially low power. Thus these tests should be used with great caution, or avoided completely. Unlike the pooled method, however, in variance weighting one does not need to assume that the rates themselves are identical across studies. Assuming that the rate of events remains stable over time, longer studies will on average have more events. Since this increase in numbers of events applies to both the PTCA and CABG groups, however, the relative risk should not be affected, and so RR's from studies with different follow-up times can still be combined. Note that the weaker assumptions of the variance weighted method compared to the pooled method results in a wider confidence interval for the variance weighted method. This is a common phenomenon in statistics, where the relaxing of assumptions leads to less certainty in the estimated results. Of course, if the assumption does not hold in the first place (as is certainly the case in the pooled analyses in our example), the entire analyses that depends on that assumption is not valid.

3.3. Meta-Analysis Using a Fixed Effects Bayesian Approach

Recall from Chapter 2 that in a Bayesian approach, one combines prior information (in the prior distribution) with the information in the data (through the likelihood function) to derive the posterior distribution. The posterior distribution summarizes all of the information that was known about the effect of interest a priori together with the additional information contained in the data being analyzed.

A very simple Bayesian model can be formed by following the idea from the pooled analysis discussed above. In each treatment arm, the total numbers of events can be considered to follow independent binomial distributions. If the prior information about each event rate can be summarized in by beta densities, the posterior distributions also follow a beta densities (see Chapter 2, Section 9.1). Roughly speaking, the ratio of these two beta densities provides the posterior distribution for the relative risk, from which a 95% credible interval can be formed. In our example from Table 7.1, the binomial likelihood functions would be:

$$\theta_1^{136}(1-\theta_1)^{(1710-136)}$$

and

$$\theta_2^{127}(1-\theta_2)^{(1661-127)}$$

where θ_1 and θ_2 are the event rates in the PTCA and CABG groups, respectively. If a diffuse or noninformative beta(0.5, 0.5) prior density is used for each group, the posterior densities are the same as the likelihood functions above, except that 0.5 is added to each exponent (four times). Thus the posterior density for θ_1 is a beta(137.5, 1575.5) density, and the posterior density for θ_2 is a beta(128.5, 1535.5) density. These densities are depicted in Figure 7.1, where one can see that the two event rates are very similar, and both are almost surely between 6% and 10%.

Technically, using Bayes Theorem to derive the posterior distribution for the RR involves methods from calculus, including a change of variable transformation and integration, so the details will be omitted here. A simple way to conceptualize the analysis is to imagine that first a large random sample is taken from each of the two beta densities, representing the rates from each treatment arm, and a new random sample is created from these by forming the ratios of the randomly selected rates in the two groups. This random sample can be considered as having arisen from the posterior distribution of the relative risk, so that a histogram of the variables generated in this way approximates the posterior density curve. Either way, the results should be very similar to those presented in Table 7.2 for the fixed effects Bayesian approach, with a point estimate for the relative risk of 1.04 and a 95% credible interval of (0.83, 1.31). Note that this interval is very similar the previous pooled estimate. Figure 7.2 displays the posterior density for the RR. One of the advantages of the Bayesian approach is the ability to directly calculate and report probabilities relating to the RR. For example, if the region of clinical equivalence is determined to be (0.9, 1.1), representing a 10% difference in rates in either direction, then the area under the curve (Figure 7.2) between these two points provides the probability of clinical equivalence of these two treatments. Here we find that $Pr\{0.9 < RR < 1.1\} = 0.57$. Therefore, we are 57% certain that these two treatments are clinically equivalent.

Of course, this very simple Bayesian model suffers from the same drawbacks as the frequentist pooled estimate discussed above, and thus cannot be realistically applied to our cardiology example. Nevertheless, the Bayesian approach does retain all of its usual advantages, including the possibility of formally including prior information into the analysis, including a range of prior distributions (see Chapter 2, Section 9), and ease of interpretation of credible intervals compared to confidence intervals. In addition, the Bayesian approach offers a way of evaluating the possible effects of publication bias as a partial solution to the "file drawer problem". After having analyzed the data with a "flat" prior distribution, to see the information that the data themselves provide, one can ask the following question: How strong would my prior distribution need to be in order to change the conclusions that I have formed from this meta-analysis? For example, if we add a study with 200 subjects and with an observed relative risk of 2 in favor of PTCA, say, does this change our conclusions? Such a study might have an event rate of 10% in the PTCA group, and a 20% rate in the CABG group. In other words, if there were an unpublished or undiscovered study with 20 events in 100 patients in the CABG group and 10 events in 100 patients in the PTCA group, would this be enough to change our main conclusions from the meta-analysis? Redoing the analysis with this added study as our "prior information" (thus using prior densities of beta(20, 80) and beta(10, 90) for the CABG and PTCA groups, respectively) finds the results moved to RR = 0.97 (95% CI (0.78, 1.21)), which is not a large change from the previous estimate. However, if it is possible that a similar "undiscovered" study could have

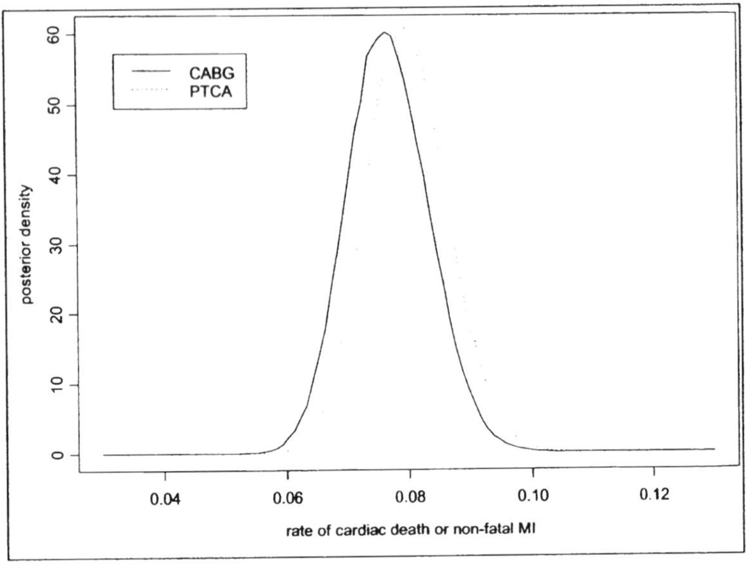

Fig. 7.1. Posterior densities for the rate of the combined endpoint of cardiac death and nonfatal myocardial infarction for the CABG and PTCA groups, for all data combined via a fixed effects effects Bayesian model.

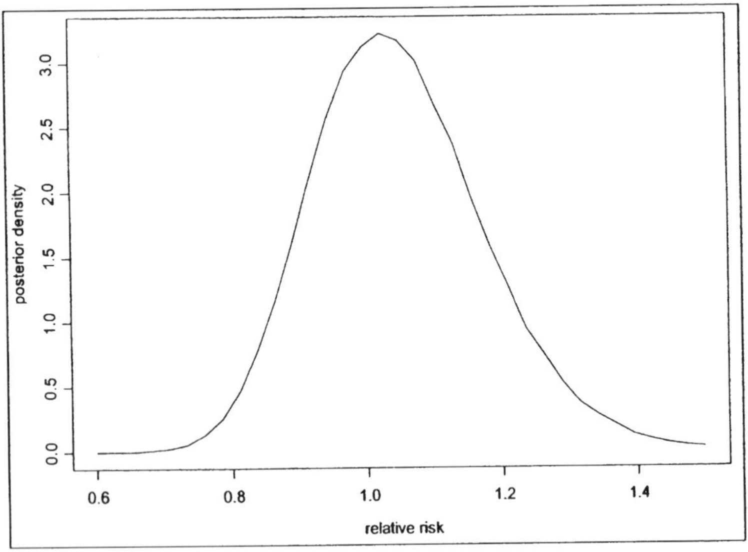

Fig. 7.2. Posterior density for the relative risk of PTCA versus CABG for the combined endpoint of cardiac death and nonfatal myocardial infarction for the CABG and PTCA groups, for all data combined via a simple Bayesian model.

had a size of 1000 rather than 200, we find RR = 0.71 (95% CI (0.60, 0.83)), which would be a strong result in favor of PTCA. Therefore, by this "backwards Bayes" technique (so-called because the prior distribution is not formed before, but rather after the data are collected), we can claim robustness to small undiscovered studies, but not to large ones. If it is considered very unlikely that such large studies would go unreported, then this analysis could show that our results are "robust" to publication bias or the file drawer problem.

4. Random Effects Meta-Analysis Models

As discussed in the previous section, fixed effects models are, at best, very rough approximations to reality. Again, this is because we rarely expect two studies to have exactly the same true effect. Variations in population characteristics, interventions applied, study protocols, and other factors mean that the effects being estimated in each study usually vary from study to study. Random effects models are specifically designed to accommodate variations in effects between studies. The basic idea is to include two variance terms in the model. The first variance term represents the usual random variability about the observed results within each study. The second variance term represents the between study variations in true effect rates (or other outcome). By considering both of these terms, one models both within and between study sources of variation. In this section we will examine two random effects models, one from each of the Bayesian and frequentist schools. As our goal is to present the basic ideas behind random effects models, technical details are left to a minimum.

4.1. Bayesian Hierarchical Random Effects Models

First consider a single study, for example the CABRI trial, whose data are given in Table 7.1. Within this single trial, we can apply simple methods to estimate the RR from this study alone. From the Bayesian viewpoint, this means we can start with two beta prior densities on the binomial probabilities for the event rates in the CABG and PTCA groups. Adding in the information provided by the data from this trial and using the methods briefly hinted at in Section 3.3, a posterior distribution for the RR for this study can be derived. Similarly, we can create posterior densities for each of the RR's from each of the eight studies considered by Pocock et al[2] listed in Table 7.1. Thus we consider the set of relative risks from each study, which we can label as RR_1, RR_2, …, RR_8.

Now, suppose that, for the various reasons mentioned above, these RR's are not all identical to each other, but rather are distributed according to:

$$RR_i \sim N(\mu, \sigma^2),\ i = 1,2,\ldots,8. \qquad (2)$$

Thus we assume that the relative risks from the 8 studies follow a normal distribution. The parameter μ represents the overall mean value, or the true average effect among all the effects in such studies. The variance parameter, σ^2 represents the study to study variability in RR's, due to the various settings of each study. Now, if $\sigma^2 = 0$, then all RR_i's are assumed to have the same true value, given by μ, and we are back to a fixed effects model. Thus we can see that a fixed effects model is really just a special case of the random effects model, when $\sigma^2 = 0$. If σ^2 is greater than 0, however, the studies do vary in their true effects, and a hierarchical random effects model is indicated.

Table 7.1. Summary of the results of the 8 trials of PTCA versus CABG as given by Pocock et al.[2] Data from Pocock et al.

Trial	Patients		Deaths or MI (%)		Mean follow-up	RR	(95% CI)
	CABG	PTCA	CABG	PTCA			
CABRI Trial[3]	513	541	29 (5.7)	43 (7.9)	1.0	1.44	(0.89, 2.22)
RITA Trial[4]	501	510	31 (6.2)	34 (6.7)	4.7	1.11	(0.67, 1.73)
King et al[5]	194	198	33 (17.0)	24 (12.1)	3.0	0.74	(0.44, 1.16)
Hamm et al[6]	177	182	18 (10.2)	10 (5.5)	1.0	0.58	(0.26, 1.14)
Puel et al[7]	76	76	6 (7.9)	6 (7.9)	2.8	1.17	(0.34, 2.96)
Hueb et al[8]	70	72	1 (1.4)	5 (6.9)	3.2	8.16	(0.60, 36.89)
Goy et al[9]	66	68	2 (3.0)	6 (8.8)	3.2	3.95	(0.61, 13.63)
Rodriguez et al[10]	64	63	7 (10.9)	8 (12.7)	3.8	1.31	(0.45, 3.01)

7

Hierarchical models such as these are often called two stage models. At the first stage, the relative risks (or other outcome of interest) have variabilities that are largely determined by the study specific data, while at the second stage these study specific parameters themselves follow a distribution that applies between studies. Thus equation (2) represents the second stage of this hierarchical model. Hierarchical models allow for the "borrowing of strength" between studies, in the following sense. Studies with large sample sizes tend to have more stable estimates compared to studies with smaller sizes. Because equation (2) acts as a sort of "meta-prior" over the collection of relative risks among the studies, μ will depend more on, and tend to be closer to, the relative risks from the larger studies. Thus small studies will tend to have posterior estimates for their relative risks that are slightly "pulled towards" the overall estimate μ. This is similar to the effect of the prior density when the data set is small in Chapter 2, Section 9.2.

In a hierarchical analysis, one can present two different types of results. First, one can present the posterior distribution for the overall mean, μ. This posterior distribution is interpreted as describing the uncertainty in the overall mean effect among the studies. Looking at Table 7.2, we see that the overall estimate of the mean RR for the 8 studies from Table 7.1 is 1.05, with a 95% credible interval of (0.74, 1.44). Notice that this is slightly wider than the previous confidence interval from the simple Bayesian estimate, because we no longer assume that the RR is constant across all studies, as in the simple Bayesian pooled approach (in fact, that approach went further, and assumed that the event rates in each trials arm were identical across all studies). Considering the entire distribution $N(\mu,\sigma^2)$ allows one to include the study to study variability due to the different settings. Thus we are able to derive the posterior density for the "next similar study" that might be performed, assuming

Table 7.2. Meta-analysis Results. Summary of the results of the meta-analysis of the data from in Table 7.1, as given by five different meta-analytic techniques. Descriptions of each technique are as given in the text. NA indicates non-applicable, as only random effects models have results for the "next study".

Method	Results for the mean effect		Results for the "next study"	
	RR	95% CI	RR	95% CI
Pooled	1.04	(0.83, 1.32)	NA	NA
Variance weighted	0.88	(0.42, 1.34)	NA	NA
Mantel Haenszel	1.05	(0.62, 1.77)	NA	NA
Simple Bayesian	1.03	(0.81, 1.30)	NA	NA
Hierarchical Bayesian	1.05	(0.74, 1.44)	1.09	(0.55, 1.97)
Dersimonian and Laird	1.02	(0.74, 1.41)	1.02	(0.56, 1.88)

that the true parameter value is a random draw from the normal distribution presented in equation (2). Here we try to capture what might occur in a randomly selected clinical setting that is subject to similar sources of variation in effects as those studies included in the meta-analysis. Assuming that the study settings included in the meta-analysis represent a reasonable range of all possible settings, this represents a more realistic assessment of what would happen in real clinical practice. Again referring to Table 7.2, the estimate of 1.09 (95% CI 0.55, 1.97) for the "next study" using a Bayesian hierarchical approach is even wider than that for the mean effect. This result recognizes that not all study settings would have relative risks near the overall mean μ. The difference between the results for the mean effect versus the results for the "next study" are analogous to the difference between reporting a standard error or a standard deviation for a mean (see Chapter 2).

Figure 7.3 presents the posterior densities for the three Bayesian approaches. As can be expected, the fixed effects Bayesian estimate provides the narrowest posterior density, although it is almost surely making unrealistic assumptions. The posterior distribution for the mean effect shows that we are quite certain that the overall mean effect is between about 0.7 and 1.5, but the final density shows that we have not ruled out substantial variation in relative risks from setting to setting.

More sophisticated Bayesian models may try to explain these differences by forming another level in the hierarchy where the mean μ is not considered as fixed, but may vary in a regression model depending on study specific covariates. See Brophy and Joseph[4] for an example of such a hierarchical model. It is important to emphasize that random effects models are appropriate when the sources of variations in the effects from study to study are unknown or uncertain. If the variations arise from well identified sources (for example, studies with different drug dosages that produce dose-effects responses), then these sources should be identified and incorporated into a more complex model, rather than considered as random effects.

Finally, as always in a Bayesian approach, prior densities are needed for the parameters μ and σ. If there is substantial prior information and the studies are either few in number or small in size, then prior information can be very useful, especially if care is taken in their elicitation and the results are presented across a

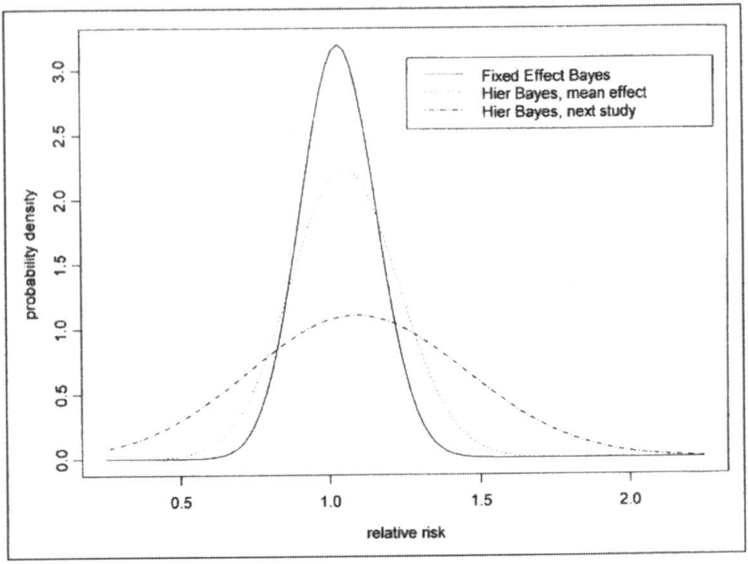

Fig. 7.3. Posterior density for the relative risk of PTCA versus CABG for the combined endpoint of cardiac death and nonfatal myocardial infarction for the CABG and PTCA groups, for all data combined from two different Bayesian models.

reasonable range of prior densities. Most meta-analyses use "noninformative" priors, however, so that subjective input is kept to a bare minimum.

4.2. The Random Effects Method of DerSimonian and Laird

Dersimonian and Laird[5] presented a frequentist method for random effects in meta-analysis. Similar to the Bayesian approach, the basic idea is to hypothesize both within and between study variances, such that

total variance = within study variance + between study variance. (3)

One then estimates the overall effect and the two variances on the right hand side of equation (3). We omit the lengthy details of the estimation procedure here, see Dersimonian and Laird[5] for the full details.

In practice, the frequentist random effect model often provides similar inferences to those from the hierarchical Bayesian approach, as can be seen from Table 7.2, where the estimated RR's and the 95% CI's are very similar between these two methods.

Random effects models, both frequentist and Bayesian, can be criticized for making unverifiable assumptions. It is typically assumed that the between study effects follow a normal distribution. Unless there is a large number of studies, the distribution of study effects is difficult to verify. While one can conceptualize using hierarchical distributions other than the Normal, this is rarely done in practice, and any choice would still suffer from the same difficulties of model verification. Furthermore, when making inferences about the "next study" setting, one assumes that the "next study" is similar to those included in the meta-analysis, which may also be either unverifiable or even unlikely, depending on the particular circumstances. Thus while random effects models are often more plausible than fixed effects models, they too suffer from great uncertainty about their applicability to many situations.

5. Summary of the Application

Considering the totality of the results presented in Table 7.2, one finds at least slightly different inferences from the different methods, even within classes of fixed and random effects models. Furthermore, Pocock et al[2] used a fixed effects model on the logarithm of the relative risk, finding estimating RR = 1.10, with 95% CI = (0.89, 1.37), which is again slightly different from any result in Table 7.2. Because they are formed from the ratios of two probabilities, the logarithms of the relative risks may be closer to normality than the relative risks themselves. Looking at the death rate alone, Pocock et al[2] report a similar result as for the combined endpoint, with RR = 1.08 and 95% CI = (0.79, 1.50). They further reported potentially important differences in the rates of revascularization and relief of angina. Thus a further difficulty with meta-analysis is that the results can depend on the specific method used to combine the data, and, of course, on the choice of endpoint.

Nevertheless, all six methods agree that the null value of RR = 1 cannot be ruled out. Furthermore, the random effects models agree that effects of up to almost 2-fold differences in event rates in either direction cannot be ruled out, at least in some settings. Thus, while no strong evidence is found for differences in the rates of cardiac deaths or nonfatal myocardial infarction, the data also do not support a strong conclusion of no difference in event rates either. Despite combining data on a total of over 3,000 patients, this meta-analysis must be considered as inconclusive, and further evidence should be gathered. Both between study and within study variability contributes to our uncertainty. Pocock et al[2] discuss many other limitations to their study.

6. Conclusion

All meta-analyses require a variety of expertises, which must be assembled before embarking on the analysis. Clinicians very familiar with the substantive area are clearly important, but so are epidemiologists who must carefully consider each study for possible biases, design flaws and other differences that may create special problems for combining study results. Statisticians should be available to provide advice on selection of the appropriate techniques to use, how to adjust for biases, if necessary, and so on.

This Chapter has presented the basic issues and simple analytic techniques behind meta-analysis. Clearly meta-analysis is a complex topic, both clinically and methodologically, and this Chapter should be considered as only a very brief introduction. Several textbooks on meta-analysis have been written which should be consulted for other methods of meta-analysis and for further details about issues related to performing and interpreting meta-analyses. Hedges and Olkin[6] is a classic text on the subject, although many new analytic techniques have appeared since its' publication. Eddy et al[7] take a Bayesian approach to meta-analysis, and include discussions of bias adjustments, combining randomized with nonrandomized studies, adjusting for differing lengths of follow-up, and other more complex topics. Cooper and

Hedges[3] is a comprehensive modern textbook on the subject that includes not only the relevant statistical techniques, but also extensive Chapters on selecting research questions, searching the literature, judging the quality of the research, and reporting the results of meta-analyses. The recent book edited by Berry and Stangl[8] contains many examples of complex meta-analyses, and discusses the bridge between meta-analyses and health policy decisions. A variety of software packages for meta-analysis are also available, such as the FastPro software of Eddy et al.[7] Some standard statistical packages also include meta-analytic techniques.

Some authors[9,10] have raised the question of whether meta-analysis has anything to offer over and above what can be concluded from a nonquantitative critical review of the literature. Clearly meta-analysis would have something important to offer IF all of the assumptions of either a fixed or a random effects model were perfectly correct, so that the answer to this question hinges on whether these assumptions are reasonable or not, and if not, on the robustness of the conclusions to deviations from the assumptions. The fact that these assumptions can be very difficult or even impossible to verify in most cases is at the root of the controversy that surrounds the usefulness of meta-analysis. Does meta-analysis provide the best possible summary of the available evidence, or does it provide an overly simplistic estimate of the uncertainty surrounding a given medical question, therefore providing a false sense of security? The jury is still out, but the answer probably lies somewhere in between these extremes for most problems.

Selected Readings

1. Last J. A dictionary of Epidemiology (Second Edition). Oxford University Press, Oxford, 1988.
2. Pocock S, Henderson R, Rickards A et al. Meta-analysis of randomised trials comparing coronary angioplasty with bypass surgery. Lancet 1995; 346(8984): 1184-1189.
3. Cooper H, Hedges L, eds. The handbook of research synthesis. New York: Sage 1994.
4. Brophy J, Joseph L, Theroux P, on behalf of the Quebec Acute Coronary Care Working Group. Medical decision making about the choice of thrombolytic agent for Acute Myocardial Infarction. Medical Decision Making 1999; 19(4):411-418.
5. DerSimonian R, Laird N. Meta-analysis in clinical trials. Controlled Clinical Trials 1986; 7:177-188.
6. Hedges L, Olkin I. Statistical methods for meta-analysis. San Diego: Academic Press 1985.
7. Eddy D, Hasselblad V, Shachter R. Meta-Analysis by the Confidence Profile Method. Boston: Academic Press 1991.
8. Berry D, Stangl D, eds. Meta-Analysis in Medicine and Health Policy. New York: Marcel Dekker 2000.
9. Bailar J 3rd. The practice of meta-analysis. J Clin Epidemiol 1995; 48(1):149-57.
10. Bailar J 3rd. The promise and problems of meta-analysis. N Eng J Med 1997; 337:559-561.

n Introduction to Decision Analysis

Alan Barkun, Neena Abraham and Lawrence Joseph

Clinical Judgement and the Science of Decision Making

Most physicians make dozens of decisions every day. Each decision has the tential to impact on that physician's subsequent decisions, influence other decins made by the members of the health care provider team and ultimately, affect tients. What shapes the medical decision-making process? The strategy most often ployed by practitioners is a vaguely defined distillation of existing knowledge, nical teaching, previous experience, instinctive choices and educated guessing. ch of these sources of information may play some role in the clinician's thought ocess, eventually leading to a decision or a specific course of action. Thus, the cision making process commonly used to guide clinical care is mostly ad hoc and ormal.[1] For simple problems, or when all of the alternatives will lead to similar tcomes, this approach may be sufficient. However, the human ability to think in ny dimensions simultaneously is limited, so the potential for poor decisions in mplex problems is not negligible.

Faced sometimes with a plethora and at other times a paucity of evidence, and th a myriad of clinical consequences to consider, generating a sound evidencesed decision can be overwhelming for the clinician. In this Chapter we discuss a mal systematic approach by which one can evaluate many clinical questions. This proach is referred to as decision analysis.

In the upcoming sections of this Chapter we describe and discuss the basic produres involved in a decision analysis, including decision tree construction, elicitan of probabilities for events in the tree, utility scores for clinical outcomes, t-effectiveness data and sensitivity analysis to all of these inputs. Throughout, we strate the basic steps involved in a decision analysis via an example relating to isions in the management of patients with symptomatic cholelithiasis and sused common bile duct (CBD) stones. This example is simplified in order to not the reader in the details of the many branches (sometimes thousands of branches included in a decision tree) that would be needed to fully model the clinical ation realistically. Nevertheless, all important concepts and steps involved in a ision analysis are included in this example, so that more complicated clinical isions can be analyzed by applying the same steps to trees with more branches. ce the basic concepts are understood and the decision tree and other inputs given, becomes a bookkeeping exercise that computer programs such as DATA (TreeAge tware Inc., Boston MA) or SMLTREE (J. P. Hollenberg, Roslyn, NY) can do

gical Arithmetic: Statistics in Surgery, by Lawrence Rosenberg, Lawrence Joseph, Alan Barkun. ©2000 Landes Bioscience.

automatically. Therefore, after reading this Chapter, you should be able to understand how to create and interpret not only simple decision problems, but also more complex trees that rely on the same principles.

Introduction to the Basic Principles of Decision Making

Decision analysis can be defined as a systematic approach to decision making under conditions of uncertainty.[1] The aim of this Chapter is to provide the clinician with basic insight into the creation of a decision model so as to interpret more easily decision analyses published in the literature or to begin to undertake such analyses with the help of methodologists.

It is important to note at the outset that the validity of a decision analysis is dependent on the accuracy of the probability estimates and the clinical relevance of assumptions utilized in constructing the decision model. These in turn are dependent on available literature and/or the opinions of an expert panel.

The implementation of a decision analysis can be sub-divided into six steps[2] as shown in Figure 8.1.

Traditionally, a clinical approach is defined by a specific set of tests and procedures used in carrying out diagnosis and treatment of patients. Each approach is visually represented by part of a decision tree, which is read from left to right, starting at a decision node. A decision node is a bifurcation point which represents the original clinical decision or health state being considered and from which all subsequent management alternatives follow. From this node, smaller branches emanate denoting chance events, further decisions, and end-of-branch (terminal) nodes with utilities representing the value or desirability of the experience of a patient who undergoes the events leading to that node. Notationally, squares usually represent decision nodes, and circles represent chance occurrences to which probability assumptions may be attributed. The complexity of the model adopted varies from simple decision paths to more complicated management schemes that attempt to emulate all aspects of clinical decision making.

According to Figure 8.1, the first step is to map all the pertinent courses of action as they apply to the treatment decisions being compared, and determine their clinical consequences. Figure 8.2 provides a simple example. In the present case, a surgeon is planning to carry out a laparoscopic cholecystectomy on a patient with symptomatic cholelithiasis. The patient displays no history of jaundice, cholangitis or pancreatitis, and has normal liver and pancreatic tests. Furthermore, an abdominal ultrasound demonstrates no biliary anomaly except for the presence of gallbladder stones. Thus the patient is, by all standards, at low risk for carrying a CBD stone. The surgeon is considering whether to perform an intraoperative cholangiogram at the time of laparoscopic cholecystectomy or not, i.e., in this case, whether to adopt a policy of routine versus selective cholangiography. This decision is represented by the initial decision node, at the left of the decision tree. For purposes of demonstration, the tree chosen represents a simplified schema not detailing in its structure laparoscopic conversion rates to open surgery, false positive or false negative cholangiograms, or failure of laparoscopic treatment, and grouping together procedural related and disease related complications. Moreover, death is not considered in this example.

1. Determine the question to be answered and consider all treatment alternatives.
2. Define the decision tree and assign probabilities to each branch point.
3. Define a utility score for each potential outcome.
4. Combine the probabilities and utilities for each node on the decision tree with the technique of "backfolding the tree."
5. Choose the set of decisions that lead to the highest expected utility.
6. Sensitivity analysis: Compare outcomes as each input is varied over its range of potential uncertainty. Vary each input separately and in combinations in order to evaluate the worst and best case scenarios and to establish the critical factors in the decision model.

Fig. 8.1. The six steps involved in a decision analysis.

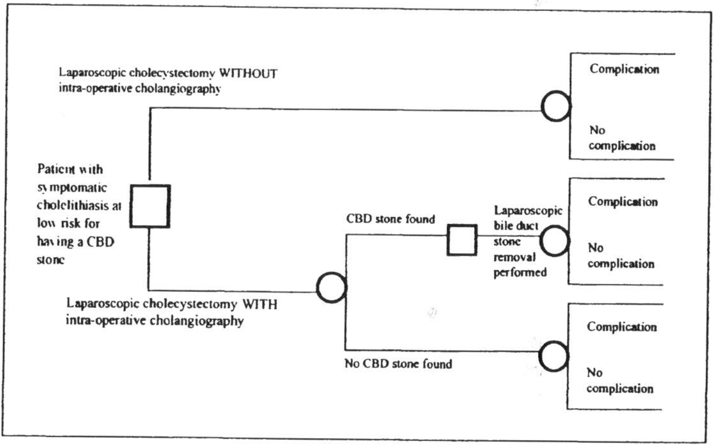

Fig. 8.2. Schematic diagram of a simplified decision tree.

As you follow along this figure from the initial decision node on the extreme left side, you can track the health states being considered. Initially, a decision is made as to whether an intra-operative cholangiogram at laparoscopic cholecystectomy is performed or not. It is this decision that we would like to evaluate, the main question being: Of these two clinical paths, is there one path which, on average, leads to better clinical outcomes? If an intra-operative cholangiogram is carried out, there are two possible outcomes: a stone can be found, or no stone is detected. These are determined according to probabilities, not based on a clinical decision, and are thus represented as chance nodes (open circles). If a stone is found, in this case, laparoscopic management with bile duct exploration and stone removal is chosen and carried out. Note that this is a decision node from which other alternative treatment strategies such as postoperative endoscopic retrograde cholangio-pancreatography (ERCP) could arise if clinically relevant as the structure of the tree is made to become more complex. All terminal health states are subsequently shown, following chance nodes, as resulting from a complication or an uncomplicated course.

In the second step, probabilities are assigned to each branch point, emanating from each of the different chance nodes. For each branch point, probability values are established based on the existing literature, and /or expert opinion.[3] For example, in Figure 8.3 probabilities are listed for each of the possible chance nodes.

The probabilities shown above were derived from the literature and an expert consensus panel and were slightly modified from a decision model carried out by our group.[3] As in many decision models, the literature does not provide for exact point estimates of every probability for every chance node in the model. In this case, for example, the proportion of retained stones that become symptomatic is largely unknown (especially when varying assumptions about the patient population) and can only be extrapolated indirectly from other similar questions addressed in the literature or expert opinion. Nonetheless, the probabilities need be internally consistent, and note that for any given chance node branching, the probabilities for each branch at that chance node are P and 1-P respectively (the sum of P and 1-P always being 1).

8

The next step in developing the decision tree is the assignment of effectiveness scores that reflects the clinical benefit or disbenefit related to the procedures and outcomes that lead to each terminal node. Often a utility score is attributed to each potential terminal node outcome. A "utility score" is similar to a "value" or "worth" estimate. Some health states are considered more desirable by patients and clinicians than others in terms of the quality and quantity of life. While often difficult to quantify, the combination of quality and quantity leads to a numerical "utility".[4] For each terminal node in the tree, there is a corresponding utility score. In order to quantify utility scores, a common scale is required. Where an obvious numerical value does not exist in the literature for a utility score, an ad hoc procedure could be to rank the outcomes from best to worst. A numerical value is then assigned to each rank,[2] such that a utility of zero would be assigned to the least desirable outcome, and a value of one would be representative of the most desirable outcome. This has the disadvantage of assuming equal step sizes between adjacent outcomes, so should be used with caution and as a last resort only.

The quantification of utility scores, in the absence of support from the literature, relies on expert opinion from patients, clinicians and researchers who estimate the value of the outcomes under investigation. Alternately, the unit of effectiveness adopted for each terminal health state may be tailored to the given clinical question, and available data from the literature. In the present example, duration of hospital stay was chosen as it is readily available from the literature and clinically most meaningful. No score was attributed to the health state death as it is a rare event with an incidence under 1/3000 in unselected series of laparoscopic cholecystectomy, and because more detailed modeling has suggested little difference with regards to the number of deaths when assessing different management strategies in patients undergoing laparoscopic cholecystectomy.[3] Furthermore, in this example, no morbidity or prolongation of hospital stay was assigned to intra-operative cholangiography (i.e., false positive and negative cholangiograms are not considered here). The values given in Table 8.1 are modified from a more detailed decision model.[3]

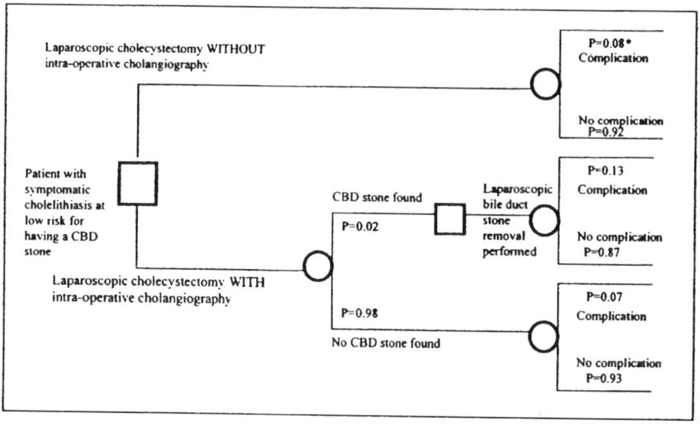

Fig. 8.3. Assigning probabilities for each branch point. The asterisk (*) indicates that the complication rate of 8% is calculated as a baseline rate of 7% (similar to the lowest branch of the tree, when no CBD stone is found), and a 50% complication rate in the two percent chance that a stone is present (i.e., a complication attributable to a retained stone), thus an additional 1%, for a total of 8%. As we will see later in the text, if the rate of stones increases, this complication rate will also increase.

8

The fourth step involves the combination of the probabilities and utilities for each node on the decision tree to find the average utility of each decision option. The basic idea is to choose the decision that leads to the optimal average utility (here, the minimum number of days hospitalized). Although results will vary from patient to patient, this decision will lead to the best outcomes, on average. The average utility for each decision option is calculated by "backfolding" the tree. The probability of eventually arriving at each terminal node is calculated by multiplication of the different chance node probabilities leading to this terminal health state (if events "downstream" are independent from previous events, which is often a reasonable assumption). The average utility contribution of each path is determined by multiplying the utility score for that path by the probability of arriving at that path's terminal node. Graphically, this is portrayed in Figure 8.4.

Using the example from Figure 8.3, the probability and effectiveness score (i.e., duration of hospital stay) for the terminal node of the decision tree that corresponds to a patient having an uncomplicated course following laparoscopic cholecystectomy and a negative intra-operative cholangiogram (the lower most node on the right) are: $0.93 \times 0.98 = 0.9114$ and 1.5 days, respectively. Similarly, a patient undergoing an uncomplicated laparoscopic bile duct stone removal following a positive cholangiogram would be expected to achieve this health state with a probability of $0.87 \times 0.02 = 0.0174$ with an average duration of hospitalization of 2.5 days.

In the fifth step, the investigator must choose the set of decisions that lead to the highest average utility. This is also referred to as "maximizing the expected utility". This is done after the expected utility for each branch has been calculated by multiplying the probability of each terminal branch by its corresponding utility score, as

Table 8.1. Examples of hospitalization durations for health states shown in Figure 8.3.

Clinical Event	Average Hospitalization (days)
Uncomplicated Laparoscopic Cholecystectomy	1.5
Complication from Laparoscopic Cholecystectomy, CBD stone removal or retained stone	3.8 (weighted average considering proportion and hospitalization for each)
Uncomplicated laparoscopic common bile duct exploration and stone removal	2.5

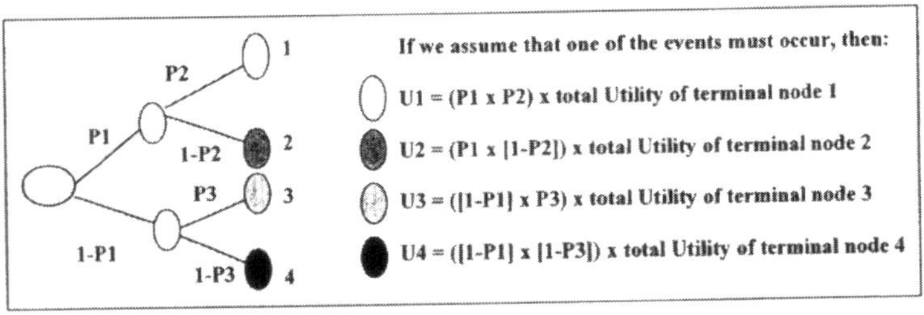

Fig. 8.4. Determination of the expected utility for each path: backfolding the tree.

depicted in Figure 8.4. In order to choose the set of decisions that leads to the highest expected utility (in our example, the lowest duration of hospital stay), one must backfold each set of outcomes and probabilities that leads to a terminal node of that approach.

Once established for each branching path, the sum of all the terminal branch utility scores emanating from each separate decision choice leads to the overall expected utility for that clinical strategy. This is depicted in Figure 8.5.

In the present example, where we only consider patients at low risk of having a CBD stone, the no cholangiography approach would include the weighted sum of the two uppermost terminal nodes of Figure 8.3. This leads to the following calculation:

$$\text{Average stay} = 0.08 \times 3.8 + 0.92 \times 1.5 = 1.684 \text{ days}$$

On the other hand, a selective cholangiographic approach would include the weighted sum of the lower 4 terminal nodes:

$$\text{Average stay} = (0.13 \times 0.02 \times 3.8) + (0.87 \times 0.02 \times 2.5) + (0.07 \times 0.98 \times 3.8) + (0.93 \times 0.98 \times 1.5) = 1.681 \text{ days}$$

In other words, for 1000 patients treated with each approach, a policy of routine cholangiography would result in a similar number of days of hospitalization (1684 versus 1681 days) as a policy of selective cholangiography (for the given set of assumptions, including a 2% prevalence of CBD stones), suggesting that the two

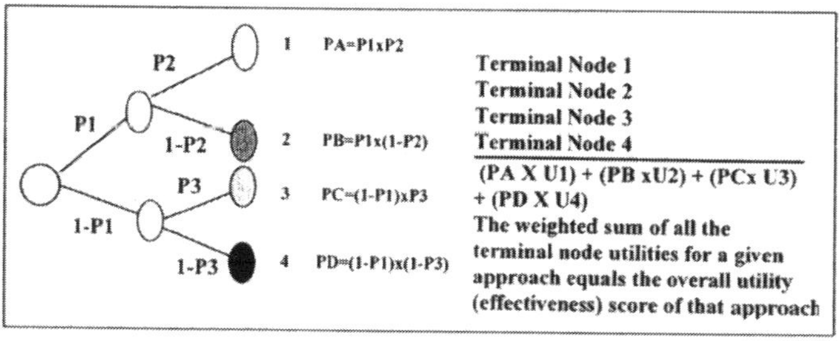

Fig. 8.5. Calculating the average utility score for a set of branches.

approaches are of approximately equal effectiveness for the measure of days hospitalized. The above results are somewhat fictitious owing to the simplifications used for the example, but provide the reader with a concrete example of the process used in calculating and analyzing the results of a decision tree.

Finally, the sixth step is designed to check the robustness of the analysis by determining its vulnerability to clinically plausible changes in the estimates of probabilities and utilities. This is also referred to as a sensitivity analysis.

Although decision analysis permits an assessment of the many different clinical approaches available in any clinical situation, the analysis is limited by the accuracy of the assumptions made in deriving the probabilities and utilities included in the model.[1] Where the literature does not provide accurate probability estimates, and where there may be uncertainty in the utilities used and the completeness and accuracy of the chosen pathways, sensitivity analysis permits inclusion of the uncertainty around the adopted point estimates and decision tree structure. Sensitivity analysis is accomplished by varying the probability estimates over a wide but realistic range of possibilities, to encompass the uncertainty inherent in the point-estimate. Often the range chosen includes a wide sample of different point estimates derived from the literature. This is especially important when considering regional differences in success rates for technical procedures, and the rapid evolution of technical equipment and expertise.[5] As the range is varied, one can determine which estimates of probabilities and effectiveness scores affect the observed results in a clinically meaningful way.

In our example, the reader will appreciate how varying the prevalence of bile duct stone can impact on the optimal clinical decision. Indeed, increasing the prevalence of bile duct stones in this population to 40% (effectively choosing a higher risk group with different clinical characteristics) provides us with a complication rate in the no cholangiography branch of $0.07 + 0.5 \times 0.4 = 0.27$, or 27%, and the probability of a stone in the lower branches of the tree increase from 0.02 to 0.4. Therefore, for 1000 patients treated, we obtain 2121 total days hospitalized if no cholangiogram is performed, as seen by calculating:

$$0.27 \times 3.8 + 0.73 \times 1.5 = 2.121.$$

In contrast, we calculate:

$$(0.13 \times 0.40 \times 3.8) + (0.87 \times 0.40 \times 2.5) + (0.07 \times 0.6 \times 3.8) + (0.93 \times 0.6 \times 1.5) = 2.064,$$

which works out to 2064 days hospitalized per 1000 patients, when performing a cholangiogram in such a population at risk for CBD stones days. The reader will appreciate the increased complication rate attributable to undetected CBD stones that has now increased with the prevalence of CBD stones. The model thus suggests a modest advantage for the approach utilizing cholangiography that results in a lesser duration of hospitalization in such a patient population. The magnitude of this advantage must be weighed in terms of a meaningful clinical benefit. Note that the overall durations of hospital stay for each approach increased (as expected) compared to the "base case" estimate (with a low bile duct stone prevalence) in both groups owing to the increased morbidity attributable to the higher background prevalence of bile duct stones (resulting in increased absolute numbers of disease and procedure related complications). Referring to the numbers given in Figure 8.3 and Table 8.1, we have assumed that all probabilities and "utilities" (hospital days) remain fixed at their previous levels, except for the probability of a bile duct stone (and the resultant complication rate in patients not undergoing cholangiography). This is called a one-way sensitivity analysis, because only one input has been changed from the "base case" tree. As another one way sensitivity analysis, suppose that we return the probability of a bile duct stone to 2%, but suppose that 5 days of hospitalization, rather than 3.8 days, are required when complications are present. Backfolding this tree results in an average days hospitalized of 1.78 in the no cholangiography group, versus 1.77 days in the cholangiography group.

Two-way sensitivity analyses can also be performed, by changing two variables at the same time. For example, we can change both the prevalence of bile duct stones (from 0.02 to 0.4, as above) and the numbers of days hospitalized for patients with complications (from 3.8 to 5, again as above). In this case, after backfolding the tree, we find that the average days hospitalized in the no cholangiography group is 2.45, versus 2.18 days in the cholangiography group. Thus if these were the true values, we would prefer the cholangiography decision.

While one-way and two-way sensitivity analyses provide some useful information, they do not expose the full uncertainty inherent in any decision analysis problem, since in reality all inputs are uncertain simultaneously. One way to address the total uncertainty is through Monte Carlo sensitivity analyses.[6] Consider Figure 8.6, which repeats the information found in Figure 8.3 and Table 8.1, but now explicitly acknowledges the uncertainty in each probability estimate and utility (days hospitalized) estimate by providing a 95% credible interval (see Chapter 2, this volume) inside of which we are 95% certain that the true value. Each of these intervals can be matched to probability distributions which are fixed such that they cover the same range, and have the same 95% intervals as desired for each uncertain parameter in the tree. The basic idea of a Monte Carlo sensitivity analysis is to incorporate all of this uncertainty simultaneously, in a single sensitivity analysis. The end result is then 95% credible intervals for the number of days hospitalized in each decision arm of the decision tree, rather than simple point estimates (see Chapter 2 for the definitions of point estimates and credible intervals). We can then not only see what the "best guess" is for each arm, in terms of average days hospitalized, but also assess how certain we are of these estimates. A Monte Carlo sensitivity analysis is easy to conceptualize, as it can be broken down into four steps:

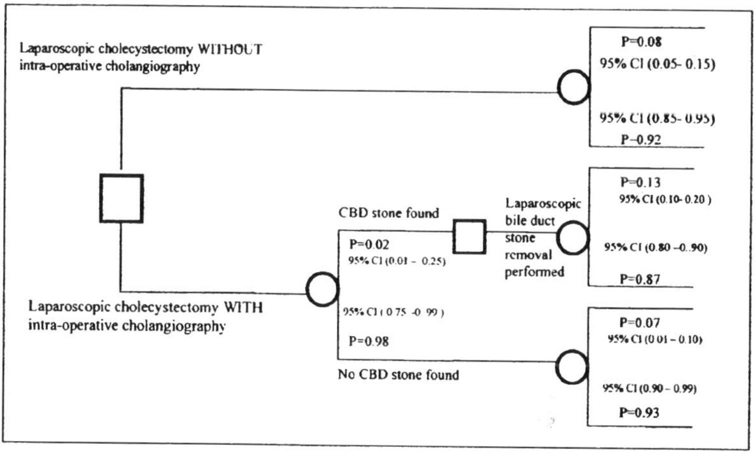

Fig. 8.6. Decision tree with probabilities and corresponding 95% confidence intervals. The 95% credible intervals for utilities that have been chosen in this example are: uncomplicated laparoscopic cholecystectomy 1.5 days (95% CI: 1-2 days), complication from laparoscopic cholecystectomy or CBD stone removal or retained stone 3.8 days (95% CI: 3-5 days), and uncomplicated laparoscopic common bile duct stone removal 2.5 days (95% CI: 2-3 days). These intervals, and those for the probabilities given in the tree itself, are not necessarily based on hard data, but will suffice for our illustrative purposes.

1. Using the distributions at each probability or terminal node in the tree, select a random value that will be used for the current iteration in the Monte Carlo analysis. [Technical Note: Here we have assumed that each probability value is selected independently from the values selected at other probability nodes, and that the utilities associated with each complicated and each noncomplicated nodes are the same, regardless of the path that lead to the terminal node. Other schemes with other levels of dependencies are also possible.]
2. Assuming that the values selected in step 1 are perfectly correct, backfold the tree as usual, and find the average days hospitalized in each arm of the decision tree. Store these values.
3. Repeat steps 1 and 2 a large number of times, typically 1,000 or 10,000 times (as you can imagine, Monte Carlo analyses are usually done by computer programs such as those mentioned earlier in this Chapter). Store each of these values.
4. Create histograms or other summary statistics (again, refer to Chapter 2 if you are not familiar with common descriptive statistics) such as means and 95% credible intervals from these stored values.

To summarize, a Monte Carlo sensitivity analysis is similar to doing, say, 1000 separate decision analyses, each with an identical (or very similar) tree structure, but with randomly varying probabilities and utilities. The output of these 1000 different analyses are then reported using, for example, 95% credible intervals for the average utility of each possible decision in the tree, or are otherwise graphically or numerically summarized.

For example, using the values in Figure 8.6, we find that a 95% interval for the number of days hospitalized in the no cholangiography group goes from 1.26 to 2.22, while a 95% interval for the cholangiography goes from 1.32 to 2.23. Note that these two intervals are very close to each other, driving home the fact that we are quite uncertain about the clear superiority of one approach over the other for the given assumptions. In fact, we can also calculate the proportion of the iterations in which the no cholangiography group had a lower average number of days compared to the cholangiography group. Doing this calculation here, we find that we prefer not to perform a cholangiography for 66% of the possible parameter sets that were drawn, meaning, roughly, that we are only 66% sure that this is the better option.

As can be seen, Monte Carlo sensitivity analysis is a simple but very powerful technique, and most decision analyses are incomplete without one.

Incorporating Cost-Effectiveness in Your Decision Analysis

In many cases, an important element to consider when creating a DTA is the inherent cost of each health strategy. Indeed, each time a physician adopts a decision, there is an attributable cost to the patient, hospital or health care system. In evaluating possible decisions, it is thus important to weigh the dollar costs to the health care system against the utility of outcomes to the individual patient or population group.

Prior to carrying out the analysis, each unit of health care utilization must be assessed in terms of the anticipated benefit and the associated costs. The individual attributable costs for each approach would include personnel, equipment, and other direct and indirect costs. One usually also allows for a discount rate which accounts for the present values of expenses to be incurred into the future (if any). Often the calculations are based on an arbitrary number of persons served in order to facilitate comparison of competing strategies. The approach with the lower cost per unit benefit would then be declared the most "cost-effective". Therefore, the total cost of each component of a given approach is summed, and its cost-effectiveness ratio is tabulated. These results are compared between the different approaches to determine the most cost-effective one. More detailed descriptions of these principals can be found in the cost-related Chapters of this book.

As a simple example, we again refer to the decision tree as represented by Figure 8.3. Since we have seen that the outcomes (days hospitalized) for the two clinical strategies in the tree are very similar, one may wish to base decisions on which treatment strategy leads to the minimum cost. Suppose that the costs associated with each terminal node are as follows (roughly based on references 7 and 8):

Cost of uncomplicated laparoscopic cholecystectomy: $8,000
Cost of uncomplicated laparoscopic common bile duct exploration: $11,000
Cost of a complication (of any type): $20,000

Here, following the same "backfolding the tree" analyses as was carried out for hospital days, but now replacing the number of days (1.5, 2.5, and 3.8) with their corresponding costs (8,000, 11,000, and 20,000, respectively), we find that the average cost for the noncholangiography decision arm is calculated as:

$$\text{average cost} = 0.08 \times 20{,}000 + 0.92 \times 8{,}000 = \$8960.00.$$

A selective cholangiographic approach would again calculate the weighted sum of the lower 4 terminal nodes of Figure 8.3, but with the above costs substituted for the days hospitalized. We calculate:

$$\text{average cost} = (0.13 \times 0.02 \times 20{,}000) + (0.87 \times 0.02 \times 11{,}000) + (0.07 \times 0.98 \times 20{,}000) + (0.93 \times 0.98 \times 8{,}000) = \$8906.60.$$

Therefore, it appears that it is cost effective (although by only \$8960.00–\$8906.60 = \$53.40) to perform a cholangiography, at least in this simplified illustrative example.

Of course, we can also perform sensitivity analyses on these costs estimates, following the same procedures as we did when the outcome was days hospitalized. In this case, we would need to vary our cost estimates within a realistic range for these costs.

The Final Analysis

Advantages of Decision Tree Analysis

The evaluation of complex clinical decisions requires a systematic approach. Decision analysis ensures logical consistency in balancing the risks and benefits inherent in any complicated decision making process. All assumptions inherent to the DTA are made explicit. Furthermore, by incorporating sensitivity analyses, DTA can be useful in identifying areas where further research is needed. Cost-effectiveness analyses arising from a DTA can be very helpful to health care planners. 8

Disadvantages of Decision Tree Analysis

Often the problems or hypotheses we wish to address are more complex than can be modeled successfully. In this case, an epidemiologist or statistician can be very helpful in suggesting alternatives to simple DTA such as influence diagrams and Markov models; alternately, one can consider a simpler tree that will hopefully contain a majority of the influential components. Another frequently cited disadvantage of DTA is the lack of sound data for good evidence-based probability and utility estimates. Subsequently, the quality of these estimates will suffer and the outcomes of the analysis will reflect the uncertainty of these suppositions.

In concluding, a note of caution: Many, if not most decision models are used to factor in the cost-effectiveness of different approaches. These cost considerations should not take priority over effectiveness performances, but as demonstrated above, both can be estimated using decision modeling. Clearly a payer perspective is adopted when assessing the results of these decision models where it is hypothesized that large numbers of patients (usually thousands) will be treated. Decision makers usually will assess these results amidst a fiscal reality of limited resources. This perspective may be quite different, however, than the one most often faced by the clinician attempting to make a decision about an individual patient.[9] This difference in perspective must be born in mind when interpreting the data from decision models and integrating the results into clinical practice. Increased familiarity with the methodology adopted will no doubt assist clinicians in this task.

While this Chapter has provided an introduction to all of the basic steps involved in a decision analysis, entire books have been written on the subject. Interested readers are referred to the books by Weinstein[1] and Smith,[10] among many others.

Selected Readings

1. The elements of clinical decision making. Chapter 1. In: Weinstein MC, Fineberg HV, Eds. Clinical Decision Analysis. Philadelphia: WB Saunders Company 1980.
2. Kassirer JP, Pauker SG. The toss-up. NEJM 1981; 305:1467.
3. Abraham AN, Barkun JS, Barkun GM, Fried LJ, and the McGill Gallstone Treatment Group. What is the optimal management of patients with suspected choledocholithiasis in the era of laparoscopic cholecystectomy?—A decision analysis. Gastroenterology 1999; 116:G0012.
4. Sackett DL, Haynes RB, Guyatt GH et al. Clinical Epidemiology a Basic Science for Clinical Medicine. Boston: 1991, 2nd Edition, P.142
5. Bouchard S, Barkun AN, Barkun JS et al. Technology assessment in laparoscopic general surgery and gastrointestinal endoscopy—science or convenience? Gastroenterology 1996; 110:915-925.
6. Doubilet P, Begg CB, Weinstein MC, et al. Probabilistic sensitivity analysis using Monte Carlo simulation. A practical approach. Medical Decision Making 1985; 5(2):157-77.
7. Liberman MA, Phillips EH, Carroll BJ et al. Cost-effective management of complicated choledocholithiasis–laparoscopic transcystic duct exploration or endoscopic sphincetrotomy. J Am Coll Surg 1996; 182:488-94.
8. Sahai A, Mauldin P, Marsi V et al. Bile duct stones and laparoscopic cholecystectomy: A decision analysis to assess the roles of intraoperative cholangiography, EUS, and ERCP. Gastrointestinal Endoscopy. 1999; 49(3):334-343.
9. Kjellstrand CM. High-technology medicine and the old: the dialysis example. J Int Med 1996; 195-210.
10. Smith JQ. Decision analysis: A Bayesian approach, Chapman Hall, London, 1987.

Pharmacoeconomics and Surgery

Pierre MacNeil and Lawrence Rosenberg

Introduction

In the last century, and especially over the past 10 years, biomedical science has ogressed rapidly. Selected cancers can be treated, organs can be transplanted, and nts can be replaced. In addition, drugs that can lower the risk of myocardial arction and prevent fractures caused by osteoporosis, are now commonly pre- ibed. The treatment options are endless, and everyday brings a new technology at promises better health. However, all new and existing treatments come at a ce; and the question of what is expensive vs. what is affordable, is one which must constantly addressed. One thing, however, is certain. After 30 years of double git growth in health care budgets, it is clear that the amount of resources that can devoted to health care is finite. With this realization, the necessity of developing ls to assist in making tough choices between interventions has emerged.

What Is Pharmacoeconomics?

Health care decision makers and health professionals have not always shown an erest in the relationship between the cost and effectiveness of health care inter- tions. In the past few years, however, it has become clear to all stakeholders in health care field, that a strategy must be developed to slow down the spiraling rease in costs. Although cutbacks and arbitrary restrictions do make it possible to ieve immediate budgetary goals, they are certainly not a long-term solution. In t, as shown by many authors, restrictions may often lead to sizable additional ts, because they only take into consideration a limited aspect of the economic pact of a given product or service.[1-4]

Economic evaluations of public services have been available since 1965,[5] and eral landmark articles suggesting the application of these technique to health care eared in the 1970s.[6,7] Today, many textbooks complement each other in provid- the neophyte with a thorough elaboration of the methods and their applica- .[8-11] An excellent book is the latest edition of Drummond,[11] for the richness of examples provided and the suggested exercises. The availability of these refer- es has coincided with a steady growth in the number of economic evaluations of th care interventions. Well over 100 studies per year are published in general lical, medical specialty, public health and policy journals.[12]

Pharmacoeconomics, is that field which explores many aspects of a given treat- it, and thus contributes to the formulation of informed decisions. It may be ned as the application of the principles of health economics to the evaluation of

ical Arithmetic: Statistics in Surgery, by Lawrence Rosenberg, Lawrence Joseph, Alan Barkun. ©2000 Landes Bioscience.

drug therapy. Pharmacoeconomic research identifies, measures and compares the costs and consequences of pharmaceutical products and services to the health care systems and society. More specifically, pharmacoeconomics can be used to do a comparative analysis of two or more choices, i.e., drug X as compared to drug Y, by examining the costs and consequences of each treatment alternative. The ultimate purpose is to ensure that optimal use is made of health care system resources, which are, by definition, scarce. In other words, the goal is to achieve the desired therapeutic objective at minimum cost, or maximize therapeutic objectives at an acceptable additional cost.

Clearly economic evaluations in health care have focused on pharmaceuticals and, hence, the emergence of pharmacoeconomics as a discipline. Although there are several reasons to explain the attention that has been focussed on pharmaceuticals (Table 9.1), the demands of an aging population, the growing constraints on health care funding, and the pervasiveness of information technology in the health care field, all conspire to make it inevitable that these evaluation techniques will be applied to all types of interventions. Surgical procedures, in particular, have already witnessed the emergence of a literature (surgicoeconomics) applying these principles of economic evaluation.[16-20]

Comparators and Perspective in a Pharmacoeconomics Analysis

9

The choice of a treatment option to which products under study will be compared is a crucial point in pharmacoeconomic analysis. If the option selected does not offer good cost-effectiveness, the other drugs being studied may be deemed useful, when in fact they are just better than a poor choice. To solve this problem, it is recommended that the most widely used drug on the market (most common clinical practice) be chosen as an option for comparison, as well as the therapy that is determined to be both effective and the least expensive[15]. In some cases, these choices will coincide.

The perspective provided by a pharmacoeconomic analysis is an important, for it determines the nature of the costs that will be included in the economic analysis. For instance, the perspective of society as a whole (all current and future members) means that all direct and indirect costs and benefits (defined in Table 9.2) must be considered.[10,21] This is the broadest possible perspective. A pharmacoeconomic analysis can also be done from other perspectives, such as hospitals, patients or third-party payers. Each perspective only considers the costs and benefits that apply to its own situation. Patients are interested in costs only to the extent that they involve out-of-pocket expenses. Yet insured patients often pay only a small portion of the actual costs. Individual patients, however, are extremely concerned with risks of prolonged disability, morbidity, mortality and quality of life. The hospital, on the other hand, is concerned only with costs during hospitalization. Effects on patient productivity and rehabilitation costs once patients are back on their feet are not included in the hospital's perspective. The narrower the perspective, the greater the chances that major costs will not be included in the analysis, and therefore, that the results will not represent the actual cost effectiveness of the products/procedure under consideration.

Table 9.1 Reasons explaining the high volume of economic evaluation of pharmaceuticals (pharmacoeconomics)

- Pharmacoeconomic evaluation are required (explicitly or implicitly) by government/formulary committees for product listing;[13-15]
- Availability of efficacy data due to the requirements for clinical trials;
- Low political liability for targeting cost-containment initiatives at pharmaceuticals;
- Availability of pharmacoeconomics data perceived as a competitive advantage within the pharmaceutical industry.

Table 9.2 Analysis of costs and consequences

	Direct	Indirect	Intangible
Costs	Professional services Drugs Laboratory tests and other examinations Medical supplies Room 7 board Home care Capital expenditures	Cost related to absenteeism Reduced productivity	Pain Mental suffering
Consequences	Changes in physical, social and mental condition Change in use of above mentioned resources	Reduced absenteeism Increased productivity	Pain Mental suffering

Costs and Consequences

Regardless of whether an economic analysis is related to health services or to some other field, it must, to be complete, take into account costs and consequences (benefits, advantages, etc.). If you prefer, it must analyze "inputs and outputs". This general principle also applies to pharmacoeconomic analysis. Using the following analogy- would you be willing to pay a specific price for a package when you don't know what it contains? Conversely, would you be willing to accept a package when you know what it contains, but not what it costs? In both cases, the relationship between cost and consequence allows us to make a decision. The concept of choice in decision-making is also very important. As mentioned earlier, health care system resources are limited, by definition. It is inevitable, therefore, that we must choose between the different ways of using these resources.

Pharmacoeconomics is a tool that can be used to clarify these choices. To effectively play this role, pharmacoeconomics attempts to make these choices clearer, by considering both the costs and the consequences.

Costs and consequences are generally divided into the following three categories: direct, indirect and intangible. Table 9.2 gives a general outline of the type of costs and the consequences in each category. This classification is based on the nature of the relationship between the cost (consequence) and the intervention. Direct costs (consequences) are directly related to the intervention. Indirect costs are more like an effect of the disease itself. The intangible costs cover certain aspects that should ideally be assessed using quality of life measurements. The important thing to remember about costs (consequences) is how they are related to the perspective of a study. When the objective of a study includes the study's perspective, the reader can determine whether all of the costs (consequences) related to this perspective have been measured.

Different Types of Pharmacoeconomic Analysis

There are four comprehensive types of economic analysis that can be used as the basis for a complete economic evaluation: cost-minimization, cost-effectiveness, cost-utility, and cost-benefit analysis. It is important to note that all types of economic evaluation use parallel methods to identify and measure costs. However, they can be differentiated by the process they use to measure, evaluate and therefore represent the consequences. Furthermore, the costs and consequences of the study vary according to the perspective used.[21]

Cost-Minimization Analysis

The cost-minimization analysis is based on the following hypothesis: the consequences of each intervention studied are assumed to be equivalent. Basically, the cost-minimization analysis asks just one question: "What costs may be incurred by the different interventions?" In a cost-minimization analysis, the (health) consequences of at least two interventions are compared and assumed to be similar.[10] It is only when a considerable amount of clinical proof shows the equivalence of the consequences that a cost-minimization analysis allows us to evaluate the cost alone of each intervention. Then, since the consequences are the same, this analysis can be used to determine which intervention will produce the desired results at the lowest cost. For examples see Box 9.1.

Cost-minimization analysis should therefore be used only when anticipated therapies are proven to have identical outcomes, but possibly different costs. Cost-minimization analysis is often seen as being incomplete since only costs are evaluated. How, then, are they different from cost analyses? In practice, when the consequences of the options are identical, there is no difference: only the costs are evaluated. However in theory, to be considered complete, an economic analysis must evaluate both consequences and costs. It is only when the consequences of the options are proven to be identical, or the difference is insignificant, that the analyst can decide not to include them in the study; the study then becomes a cost-minimization analysis. If, on the other hand, the analyst decides not to study the consequences, the study is partial: it is then a cost study.

As mentioned earlier, the principal objective of a cost-minimization analysis is to determine which intervention costs the least. Because stringent criteria are required to qualify two therapies as being equivalent, the cost-minimization analysis is not often used to evaluate medical interventions. The cost-minimization analysis is used mainly in two situations. First, it may be useful in evaluating the cost difference between different dosage forms of a drug, and second, between two generically equivalent drugs in which the outcome has been proved to be equal. The cost-minimization analysis may prove to be a practical tool when used properly. However, if there

Box 9.1 Examples of cost-minimization analysis

Einarson et al[22] did a cost-minimization analysis and a cost-effectiveness analysis. From the government's perspective, they compared four treatment options for severe psoriasis: cyclosporine A, methotrexate, etretinate and PUVA therapy. They considered four major cost categories, i.e., acquisition and administration costs, the cost of routine medical care, the cost of laboratory tests, and the cost of treating adverse effects. All of these are direct costs.

In this study, the authors present the results of the analysis without taking into account the relative efficacy of the different options (cost-minimization analysis). They then present the anticipated cost calculated with the aid of a decision tree, which allows them to consider the probability of success and recurrence.

	Cyclosporine A	Etretinate	Methotrexate	Puva
Cost-minimization	$842	$722	$1172	$752
Cost-effectiveness	$1272	$1383	$1586	$1450

This illustrates the relative effectiveness of the different therapies and their effect on costs. Indeed, a therapeutic failure often leads to additional expenditures. These expenditures are considered in the cost-effectiveness analysis, but not in the cost-minimization analysis. Indeed, if the authors of this article had stopped after the cost-minimization analysis, the study would not have identified any differences. Thus, it allows the reader to appreciate the relevance of the cost-effectiveness analysis.

In presenting the results of their cost-effectiveness study Trant et al[16] stated that no statistically significant difference was shown between stent and surgery for both acute and intermediary outcome. In such a case, the analyst, when interpreting the results, shift to a cost-minimization mode. For Trant et al[16], it meant that given that stenting was at least as effective as surgery and that surgery ($58,068) was more costly than stenting ($33,809), they had to conclude that stenting was more cost-effective than surgery.

are any doubts about the equivalence of the consequences, it is recommended that a more detailed study be done, such as a cost-effectiveness, cost-benefit or cost-utility analysis.

Cost-Effectiveness Analysis

Cost-minimization analyses are based on the hypothesis that the consequences of the interventions studied are equivalent. The limitations of the this method are obvious. In practice, we often want to compare two treatment options that do not meet this criterion. The cost-effectiveness analysis allows us to make this type of comparison. The term "cost-effectiveness" is often used incorrectly and is often used to describe all types of pharmacoeconomic analysis. Cost-effectiveness analysis includes any study that measures consequences in natural units. It is the most common form of pharmacoeconomic analysis. Table 9.3 illustrates this with a few examples. Most of these measurements express effectiveness by taking into account a single dimension and only make it possible to draw comparisons between similar types of treatments. Box 9.2 presents an example of a cost-effectiveness analysis.

Table 9.3 Examples of effectiveness measurements expressed in natural units

Successful treatment (e.g., complications avoided, increase in vessel diameter by (50%)
Life-year-saved
Lives saved
Symptom-free days
Episode-free days

Box 9.2 Examples of cost-effectiveness analysis

Rosenberg et al[19] performed a cost-effectiveness analysis to establish if the additional cost associated with the use of octreotide to prevent complications in patients undergoing pancreatic resections was justified by a decrease in the consumption of other resources and/or an improvement in patient outcome.

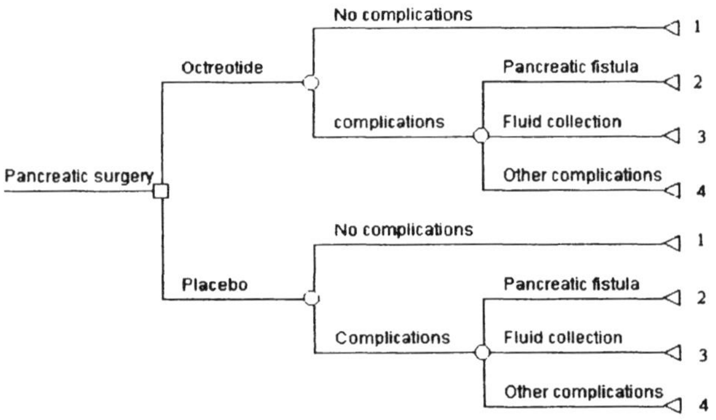

To evaluate success rates and complication rates, a meta-analysis of double-blind, randomized, controlled clinical trials was conducted. In a second phase, we evaluated the treatment cost for patients with and without complications using two costing methods. The results of these two steps where then incorporated in the decision tree depicted above.

The data suggest that when compared to placebo, octreotide is a dominant treatment strategy. In model 1, in a cohort of 100 patients, it saved an average of $853 per patient while allowing 16 incremental patients to avoid complications. In model 2, it saved an average of $1,642 per patient while still allowing 16 patients to avoid complications.

Detailed one-way and two-way sensitivity analysis suggested that both models were robust. We concluded that the use of octreotide is a cost-effective strategy in patients undergoing elective pancreatic resection. Consideration should be given to extending its use to patients that are at high risk to develop a complication following pancreatic surgery.

Obtaining and Using Effectiveness Data

There are several methods used to obtain effectiveness data to be included in a cost-effectiveness analysis. The first is to establish a cost-effectiveness analysis parallel to a clinical trial. There are two main disadvantages with this approach. First, the organization and execution of a clinical trial takes a long time and is often very expensive. Furthermore, the results of clinical trials do not focus much on the use of the drug in everyday practice. The circumstances are artificially controlled and it is often difficult to distinguish between the costs associated with the use of the drug itself and those generated by the research protocol. However, if you have the time and money, the high level of internal validity generally found in clinical research will definitely ensure credibility to the effectiveness data on which your analysis is based. For a good discussion on collecting pharmacoeconomic data alongside clinical trials see Mauskopf et al.[23]

If you have neither the time nor the money to conduct a clinical trial, it is possible to use already-existing data reported by other researchers. However, it is not uncommon to find contradictory results on the effectiveness of a product or procedure in the scientific literature. The second method, meta-analysis, is a statistical method that makes it possible to calculate a weighted average of the results of several studies in terms of the size of the sample of each study. This method is, of course, much less expensive than a clinical study and can generally be conducted within a reasonable amount of time. Since it integrates the results of several studies, and therefore examines results across a set of different conditions, it is generally recognized as having good external validity. It is not, however, without its disadvantages. For instance, very specific inclusion and exclusion criteria must ensure that the targeted studies correctly measure the same variables, according to the same criteria. For a detailed discussion on meta-analysis see Chapter 7.

If it is not possible to conduct a clinical trial, and if a review of the scientific literature does not provide enough data for a meta-analysis, it is still possible to do a cost-effectiveness analysis based on hypotheses relating to effectiveness. The results of such a study should, however, be submitted to a sensitivity analysis to determine the strength of the results obtained. Regardless of how the effectiveness data is obtained, it is still important to ensure that they are applicable to one's own practice.

Presentation of the Results

Special attention must be paid to how the results of a cost-effectiveness analysis are presented. Very often, the mean cost-effectiveness ratio, i.e., the quotient obtained by dividing the cost by the number of successes, is the only result presented. When this ratio is used as a criterion, the favored option is the one that presents the lowest cost-effectiveness ratio. However, this ratio can be misleading. Let us look at a fictitious example of this type of situation. Suppose the cost-effectiveness of drug A is compared with that of drug B. Drug A makes it possible to save 10 lives, at a cost of $1,500, while drug B saves 15 lives at a cost of $3,000. The average cost-effectiveness ratio of drug A is $150 per life saved ($1,500/10) which is superior (meaning the lower the better) to drug B ($200 per life saved or $3,000/15). Thus, according to the simple average cost-effectiveness ratio, drug A is the best alternative, since it has a lower cost-effectiveness ratio. Yet few decision-makers would choose drug A over drug B, because the incremental cost-effectiveness ratio for drug B compared with drug A is ($3,000-$1,500)/(15-10) = $300 per additional life saved. This calculation shows clearly that drug B is more expensive but also more effective than drug A. If we consider that a life saved is worth $300, we can say that drug B offers a good cost-effectiveness ratio. This value judgment is therefore closely related to the measurement of effectiveness. For instance, is a bout of nausea that is relieved worth $300?

The above example above clearly illustrates that a treatment with a good cost-effectiveness ratio is not necessarily less expensive. When a treatment alternative is both more effective and more expensive, the simple average cost-effectiveness ratio can be misleading. Under such circumstances, greater insight would be provided by analyzing incremental cost-effectiveness ratios as above, to clarify the choices available to decision-makers. They can then determine whether the additional benefits obtained are worth the additional costs. Sometimes this decision is obvious. In other cases, it will be debatable. In the event that a drug is both more effective and less expensive, it will then qualify as "dominant".

To avoid the potential confusion surrounding the different types of ratios, it is recommended to also report the results of an analysis in a disaggregate form. The costs and the consequences associated with each of the alternatives studied are presented independently, leaving it to the decision-maker to interpret the results and make the relevant comparison.[13,15] When results are presented that way, the term cost-consequence analysis is used.

Cost-Utility Analysis

The limitations imposed by single dimensional units of measurement have given rise to the cost-utility analysis, which makes it possible to measure the consequences of a treatment much more comprehensively, by considering patient preferences and quality of the health outcome produced. We saw that a cost-effectiveness analysis is a type of pharmacoeconomic analysis that compares the costs of different treatment alternatives with their consequences, and measures them in terms of natural units. In addition, we saw that it is the most common and the most widely accepted form of pharmacoeconomic analysis. However, the cost-effectiveness analysis has one shortcoming: it only considers one aspect of morbidity (single dimension).

The cost-utility analysis is sometimes considered an extension of the cost-effectiveness analysis with one difference: the measurement of effectiveness takes into account mortality, as well as the effect on quality of life (multidimensional aspects of morbidity). The consequences studied incorporate the patient's perception of the impact of the intervention on his quality of life, in addition to the incidence on clinical parameters and years of life. Thus, the measurement of consequences takes into account the evolution of survival and health related quality of life.

In economic terms, "utility" refers to the "quantity" of well-being which a person can enjoy, assuming the hypothesis ceteris paribus (all else being equal), notably the consistency of prices, preferences and individual tastes. Utility therefore indicates the preferences of an individual (e.g., the desire or preference for a given condition) in financial or medical terms. Consequently, it is logical that utility varies from one person to another. The terms utility, value and preference are often used interchangeably, but in fact, there are differences. Preference is the umbrella term that describes the overall concept; utilities and values are different types of preferences and relate to how you do the measurements. What you get depends on the way the question is framed, specifically whether the outcomes in the question are certain (value) or uncertain (utility).[11]

How Is Utility Measured?

Most commonly, cost-utility analysis measures health improvement attributable to the intervention in quality-adjusted life-years (QALYs) gained. Using the same approach as for cost-effectiveness analysis, the results are expressed in terms of cost per QALY gained. As shown in Table 9.4, methods for estimating preferences used to calculate QALYs can be divided in techniques measuring directly the preferences of individuals or prescored multi-attribute health status classification systems. The three main techniques used to measure preferences are best described by Torrance.[24] However, measuring preferences for health outcomes is a very time consuming and complex task. A recent alternative that is very attractive and being widely used, is to bypass the measurement process by use of a prescored multi-attribute health status classification system. Once the utility(or preference score) is know for a certain health state, that score in then multiplied by the length of time the patient will spend in that particular health state to obtain the number of QALYs gained.

The cost-utility analysis is particularly important when:[11]

1. a complete range of health consequences is considered important;
2. quality of life is the most important consequence desired;
3. quality of life is one of the most important consequences desired;
4. both morbidity and mortality are important consequences and a common unit is sought to measure their effects;
5. the intervention must be compared to another intervention whose cost-utility ratio has already been established.

The cost-utility analysis is therefore relevant in cases where the principal consequence sought has to do with quality of life and issues relating to morbidity and mortality must be evaluated simultaneously.

Cost-Benefit Analysis

Cost-benefit analysis is a technique that has been in use for more than fifty years as a tool to aid with the decision-making process in the development of economic and social policies. It differs mainly from other types of pharmacoeconomic analysis in that it measures both costs and consequences in monetary terms. Theoretically, this should represent an advantage, since the yield of health investments can then be compared to those of the other sectors of the economy. It is also possible, among other things, to determine whether or not a given intervention in the health field represents a net gain for society—in other words, if the sum of its benefits is greater than the sum of its costs. The theoretical and practical aspects of this approach are also outlined in a number of books.[28-30]

Cost-benefit analysis, however, does have its flaws. While measuring the consequences of interventions in monetary terms seems to be expedient on a theoretical level, the fact remains that reducing the multiple aspects of any therapy to a single monetary value is not an easy task. The methods proposed to date have been widely criticized, both on ethical and methodological grounds, and this explains why this method has few proponents.

Table 9.4 Methods for estimating preferences and deriving QALYs

Direct Preference Measures	Multi-Attributes Systems
Rating scale, category scaling, visual analog scale; Standard gamble Time trade-off	Health Utilities Index (HUI)[25] Quality of Well-Being (QWB)[26] EuroQol (EQ-5D)[27]

Three principal methods are used to perform cost-benefit analyses. The first method is based on what is called the human capital approach. According to this method, an individual's contribution to society is evaluated according to his or her remuneration. The two other methods are based on observed or expressed preferences: they determine how much money people are willing to accept (in financial terms) for an increased risk or how much they are willing to pay for a particular service (willingness to pay). We will describe these methods in more detail, but we can say from the outset that the willingness to pay method seems to be the most viable.

Human Capital

The first cost-benefit studies used the human capital method. This method is based on the concept that human beings are a capital investment like any other (at least in terms of their work life) and as a result, they produce a certain number of goods and services in the course of their life. If we consider that the value of an individual's production is equivalent to his or her remuneration, the value of the consequences (benefits) of interventions designed to maintain or improve their health can be measured in terms of the production that would be lost as a result of illness. Depending on the nature of the illness, this lost productivity could be spread over several years. In such a case, the analyst should depreciate the future benefits to bring them back to their current value.

The first criticism of this method stems from the fact that it forces the analyst to set a price on human life. Many people feel that this poses a serious ethical problem. Some people would argue that human life is priceless. Most economists would answer that a price is already implicitly put on human life in a wide range of decisions related to public expenditures.[31-32] One such example would be safety standards, relating for instance to the construction of sections of highway (particularly curves) or environmental issues. Generally, the more stringent the standards, the higher the cost to implement them. When these standards are defined, a price is implicitly set on human life and on the injuries that will occur despite the establishment of these standards. A cost-benefit analysis is simply more explicit about these values.

The other criticism, which is just as legitimate as the first, is related to the method used to evaluate the value of benefits. In fact, simplicity is probably the sole advantage of using remuneration. Imperfections in the labor market, such as professional corporations, trade unions, unemployment, and the fact that such a system has difficulty accounting for people who are not employed, like stay-at-home mothers, pensioners, and people who are unemployable, are other reasons why the human capital method is outdated. Added to this is the fact that it does not account for the pain and suffering associated with illness. Thus, it is not difficult to understand why this method is no longer used in the health field.

Observed or Expressed Individual Preferences

The approach based on observing individual preferences uses observations of individual behaviors to quantify (in monetary terms) the benefits of a given medical intervention. The observation of an individual's behavior with regard to risk is a way of evaluating the implicit value accorded to health status. People regularly accept money in exchange for an increased risk of mortality or morbidity. Electricians, race car drivers and policemen, for instance, all have what can be described as high-risk occupations, for which they receive financial compensation. If we compare the additional income with a change in the risk level of an activity, it is possible to observe the value that individuals implicitly give to different health states, and ultimately, to their life. Of course, the conversion process is not simple, but different scales have been developed for this purpose.[33] In practice, however, the limited number of situations in which attitude toward risk can be observed and measured complicates the application of this method.

The limitations of the first two methods we have presented has resulted in the development of the willingness-to-pay method. According to this method, individuals are asked to indicate their preference, in monetary terms, for different specific situations. This method was first used in the environmental field.[34] In the health field, the willingness-to-pay method attempts to establish the value of a given change in health status, by asking people how much they would be willing to pay to obtain or avoid this change.

In most cases, this method operates like an auction. The participant is given a starting price, which he accepts or rejects. Depending on the response, the price is raised or lowered to arrive at the maximum price he is willing to pay. This type of evaluation is subject to a number of biases, such as the starting price set by the investigator, and other forms of conformity or strategic bias, which can be avoided or reduced by an informed researcher. Quantifying the value of health benefits in terms of willingness-to-pay raises the issue that this willingness is closely related to the participant's income. Weighting systems based on income have been proposed, but they are still under discussion. For the time being, this limitation means that this method is best suited to inexpensive therapies. Studies conducted using this method are still uncommon.[35-37] However, it is surely the most promising method and the only way of the future for cost-benefit analyses.

Other Issues Surrounding Pharmacoeconomics

Discounting

Like in any economic evaluation the time factor plays an important role in the economic evaluation of medical interventions since it is rare that costs and consequences are all achieved at the same time, especially in the case of screening or prevention programs. It is therefore essential that correct use be made of discounting procedures to make adjustments for these time differences, to obtain current and future costs and consequences in comparable units. The need for discounting has its root in the notion that even if the inflation rate was zero and there were no bank interest, people would still consider it a benefit to receive a payment earlier or to incur an expense later. Economist call this the notion of time preference.

Although economists agree that costs should be discounted, the discount rate is still a controversial subject. There is, however, some consensus on the social discount rate, which should be between 4 and 8%, or most usually 5%. The discounting of consequences (or health gains) is another controversial subject. Logically, any consequence should be discounted, including future years of life saved. Here is why. The years of life saved are discounted, not because they can be invested or generate additional years, as is the case with money, or because their value is less than in the present, but rather because they are evaluated in relation to dollars. Since the future dollar is discounted in relation to the current dollar, future years of life must also be discounted in relation to the current year. For a detailed discussion see Krahn et al.[38]

Sensitivity Analysis

Since the value of costs and consequences are likely to fluctuate and some of their components may be difficult to estimate, it is recommended that tests be conducted to determine how results or conclusions vary when specific input variables are changed. This is called a sensitivity analysis. The sensitivity analysis allows the analyst to systematically vary the value assigned to the most uncertain variables and hypotheses when doing the calculations for an economic evaluation (such as an intervention's probability of success, the cost of new pharmaceutical technologies, the discount rate), in a certain range of plausible values. If the basic conclusion remains the same when a certain variable or hypothesis is changed, the conclusion is reliable. If, on the other hand, the conclusion is "sensitive" to the change in a variable or hypothesis, research should be continued in order to obtain additional data about these variables or hypotheses. A sensitivity analysis can therefore indicate fields in which more extensive research is required. It can be a powerful tool which analysts can use to verify the strength of their conclusions, and to identify variables that are likely to change their recommendations.

The conclusions of an economic evaluation based on uncertain data and subjective values are likely to be interpreted as being too inconclusive by some, despite all possible sensitivity analyses. Critics of economic evaluations state that this uncertainty makes any attempt to quantify the consequences of medical interventions useless. However, decisions on resource allocation must still be made, and there is often no alternative but to base them on either a reliable, but imperfect analysis, or on no analysis at all. The first solution is by far the best, given the growing complexity of the issues at hand, problems with compromise, and scarcity of resources.

The Use of Modeling in Pharmacoeconomics

Pharmacoeconomic evaluation often uses various modeling techniques which have been developed in disciplines such as epidemiology, statistics, operations research, and decision science. These techniques are used mainly in two circumstances. First, where the relevant clinical trials have not been conducted or did not include the collection of economic data, then decision analytic models are used to synthesize the best available data.[39] The study presented in Box 9.2 is an example of such a model. Second, where the clinical trials measure intermediate endpoints or have only short-term follow-up, statistical model are used to extrapolate beyond trial to final endpoints such as survival.[39] The study conducted by Riviere et al is a good example of the latter.[40]

Decision Tree

The techniques of decision analysis are in common usage by practitioners of health care economic evaluation. For detailed discussion of decision analysis see Chapter 8. In summary, using decision analysis for economic evaluation proceeds by careful structuring of the problem using a decision tree—a graphic schema where we begin with the decision (e.g., treatment A or treatment B) and trace out all probable pathways and consequences (e.g., health outcomes and costs) that can arise over time.[11] The model draws data from multiple sources—mainly clinical trials for probabilities such as rates of complication, administrative data for costs, and expert physician opinion concerning treatment algorithms.

Beyond Trial Results

Clinical trials are often constrained in terms of the length of follow-up for clinical and resource consequences. Many trials of therapeutic interventions measure short-term mortality. In many situations the economic analyst will wish to extrapolate data beyond the period observed in the clinical trial. Also, the economic analyst who wishes to link the intermediate biologic endpoint of a clinical trial to final health outcomes such as mortality reduction and life expectancy needs to rely on modeling. This is because more general measures, such as life-years saved, are more relevant to economic evaluation than short-term measures such as percentage of LDL cholesterol reduction at six months.

For a good discussion on the use of modeling in pharmacoeconomics see Buxton et al[39] and O'Brien et al.[41]

Conclusion

Pharmacoeconomics evaluations are intended to be a tool to assist in the decision making more so from a public health or general population perspective then a bed side perspective. They are certainly not meant to be or replace the decision making process which should encompass a multitude of other considerations such as clinical expertise, ethics, justice, equity or politics. By providing information derived from the best possible sources at the time of the analysis, pharmacoeconomics strives to make more explicit and evidence-based the tough choices required in allocating health care resources. Ultimately, when put into practice, such an approach should allow us to achieve the best health outcomes for the most people at the lowest possible costs.

A pharmacoeconomic analysis will always be as good as the information used to produce it. If anything, the structured process of evaluating the evidence normally used in performing a pharmacoeconomic analysis, the explicit statement of assumptions and the working-out of their implications is helpful in reaching a decision. Decisions will be made about which drugs to include on the formulary or which procedure should take place. The idea here is to structure and use the available information to help in making these decisions.

Selected Readings

1. Bloom B, Jacobs J. Cost effect of restricting cost effective therapy. Medical Care 1985; 23:872-880.
2. Soumerai SB, Ross-Degnan D, Avorn J et al. Effect of Medicaid drug-payment limits on admission to hospitals and nursing homes. N Engl J Med 1991; 325:1072-77.
3. Soumerai SB, McLaughlin TJ, Ross-Degnan D et al. Effect of limiting Medicaid reimbursement benefits on the use of psychotropic agents and acute mental health services by patients with schizophrenia. N Engl J Med 1994; 331:650-55.
4. Horn SD, Sharkey PD, Gassaway J. Managed care outcomes project: Study design, baseline patient characteristics, and outcome measures. Am J Man Care 1996; 2:237-47.
5. Dorfman R. Measuring benefits of government intervention. Washington DC: Brookings Institutions, 1965.
6. Williams A. The cost of benefit approach. Br Med Bull 1974; 1:199-225.
7. Weinstein MC, Stason WB. Foundations of cost-effectiveness analysis for health and medical practice. N Engl J Med 1977; 296:716-721.
8. Luce B, Elixhauser A. eds. Standards for socioeconomic evaluation of health care product and services. 1st ed. Berlin Heidelberg: Springer-Verlag, 1990.
9. Gold MR, Siegel JE, Russell LB et al eds. Cost-effectiveness in health and medicine. New York: Oxford University Press 1996.
10. Drummond MF, Stoddart GL, Torrance GW eds. Methods for the economic evaluation of health care programs. 1st ed. Oxford: Oxford University Press, 1987.
11. Drummond MF, O'Brien B, Stoddart GL et al eds. Methods for the economic evaluation of health care programs. 2nd ed. Oxford: Oxford University Press, 1997.
12. Elixhauser A. Health care cost-benefit and cost-effectiveness analysis (CBA/CEA) from 1979 to 1990: A bibliography. Med Care 1993; 31:JS139-JS141.
13. Ontario formulary submission
14. Henry D. Economic analysis as an aid to subsidization decisions: The development of Australia's guidelines for pharmaceuticals. PharmacoEconomics 1992; 1:54-67.
15. Canadian Coordinating Office for Health Technology Assessment. Guidelines for Economic evaluation of pharmaceuticals: Canada—2nd ed. Ottawa: CCOHTA, 1997.
16. Trant CA, O'Laughlin MP, Ungerleider RM et al. Cost-effectiveness analysis of stents, balloon angioplasty, and surgery for the treatment of branch pulmonary artery stenosis. Pediatr Cardiol 1997; 18(5):339-344.
17. Saleh KJ, Wood KC, Gafni A et al. Immediate surgery versus waiting list policy in revision total hip arthroplasty: An economic evaluation. J Arthroplasty 1997; 12(1):1-10.
18. King JT, Glick HA, Mason TJ. Elective surgery for asymptomatic, unruptured, intracranial aneurysms: A cost-effectiveness analysis. J Neurosurg 1995; 83(3):403-412.
19. Rosenberg L, Mac Neil P, Turcotte L. Cost-effectiveness analysis of octreotide in prevention of complications following pancreatic resection. J Gastrointest Surgery 1999;3:225-232.
20. Van den Oever. Cost-effectiveness in surgery: The insurers' point of view. Acta Chir Belg 1995; 95(5):205-210.
21. Bootman JL, Townsend RJ, McGhan WF. Principles of pharmacoeconomics. Cincinnati: Harvey Whitney, 1991.
22. Einarson T et al. Oral treatments for severe psoriasis: government payor analysis for Canada. J Derm Treat. 1994; Vol. 5 (suppl.1): F-23-F-27.

9

23. Mauskopf J, Schulman K, Bell L et al. A strategy for collecting pharmacoeconomic data during phase II/III clinical trials. PharmacoEconomics 1996; 9(3):264-277.
24. Torrance GW. Measurement of health-state utilities for economic appraisal: A review. J Health Economics 1986; 5:1-30.
25. Feeny D, Furlong W, Boyle M et al. Multi-attribute health status classification systems: Health Utilities index PharmacoEconomics 1995; 7:490-502.
26. Kaplan RM, Anderson JP. The general health policy model: an integrated approach. In: Spilker B ed. Quality of life and pharmacoeconomics in clinical trials. 2nd ed. Philadelphia: Lippincott-Raven 1996: 309-322.
27. Dolan P, Gudex C, Kind P et al. A social tariff for EuroQol: Results from a UK general population survey. Discussion paper no. 138, Centre for Health Economics, University of York, York, 1995.
28. Pearce DW. Cost-benefit analysis. 2nd ed. London: MacMillan, 1983.
29. Mishan EJ. Cost-benefit analysis. 4th ed. London: Unwin Hyman, 1988.
30. Walshe G, Daffem P. Managing cost-benefit analysis. London: MacMillan, 1995.
31. Jones-Lee MW. The value of life. An economic analysis. London: Martin Robertson, 1976.
32. Mooney GH. The valuation of human life. London: MacMillan, 1977.
33. Mooney GH. Human life and suffering. In: Perace DW. editor. The valuation of social cost. London: George Allen & Unwin, 1978.
34. Mitchell RC, Carson RT. Using surveys to value public goods. Washington, DC: Resources for the Future, 1989.
35. Johannesson M. Jonsson B. Economic evaluation of health care: is there a role for cost benefit analysis? Health Policy 1991; 17:1-23.
36. Acton JP. Evaluating public programs to save lives: the case of heart attacks. Santa Monica: Rand, 1973.
37. Thompson MS. Willingness to pay and accept risks to cure chronic disease. Am J Public health 1986; 76:392-396.
38. Krahn M, Gafni A. Discounting in the economic evaluation of health care interventions. Medical 1993; 31: 403-418.
39. Buxton MJ, Drummond MF, Van Hout BA et al. Modelling in economic evaluation: An unavoidable fact of life. Health Economics 1997; 6:217-227.
40. Riviere M, Wang S, Leclerc C et al. Cost-effectiveness of simvastatin in the secondary prevention of coronary artery disease in Canada. Can Med Assoc J 1997; 156(7):991-7.
41. O'Brien BJ. Economic evaluation of pharmaceuticals. Frankenstein's monster or vampires of trials? Medical Care 1996; 34:DS99-DS108.

CHAPTER 10

The Costing of Medical Resources

Ralph Crott

Introduction

Costs are an integral part (i.e., the numerator) of the different types of economic evaluation methods of health care interventions (see Chapter 9, this volume).

Although there exists a conceptual difference between cost-of-illness (COI) studies and more complete economic evaluations, the problem of estimating the cost of the medical resources used is common to all. One can say the same about whether one is interested in the costing of preventive measures (i.e., vaccination programs or prophylaxis) or interventional measures such as surgery.

Typically, a cost analysis follows a sequence of steps. These steps are outlined below, and will be discussed in a step-by-step fashion in this Chapter.

1. Defining the medical intervention or technology being evaluated
2. Defining the study viewpoint (perspective) and the type of study
3. Defining the time period of coverage for the cost data collection
4. Defining the resource items to be costed
5. Collecting data on the resources being consumed
6. Estimating the unit costs of the resources used
7. Correcting for time preference

Costs or Charges?

It is important to distinguish between two different concepts of costs: financial costs (or outlays) and economic costs.

Financial costs are the financial outlays required to produce the intervention. Typically these are estimated by the charges paid by either the insurer or the Ministry of Health (MOH) to the provider for the rendering of the medical services. For example, in Canada, many medical services such as diagnostic procedures or drugs are reimbursed from a public Provincial List of negotiated reimbursement "prices", while in the United States many medical procedures and drugs are reimbursed by private insurance or by government programs such as Medicare.

Economic costs, on the other hand, are defined to be the value of forgone opportunities or opportunity costs, that is, the value of these resources if employed elsewhere. In any given situation where available resources are not unlimited, the resources used for a specific medical intervention (e.g., nursing time, operating room use, etc...) are no longer available for another intervention or program.

Surgical Arithmetic: Statistics in Surgery, by Lawrence Rosenberg, Lawrence Joseph, and Alan Barkun. ©2000 Landes Bioscience.

This can be either at the local level (e.g., in a hospital) or at a more general level such as between two health care programs at the State, Provincial or National level. For example, an active screening program for detecting osteoporosis in the population using bone densitometry might use up medical and nonmedical resources that could have been used to expand, say, emergency services in hospitals.

In a "perfect" market place the charge or price charged for goods or services would represent their true economic cost. In reality, however, medical resources are not priced at their true economic value for several reasons. First, charges or list prices are greatly influenced by bargaining between the different parties involved (insurer, government, medical associations, suppliers, etc...). Thus the list price may differ, even for the same intervention but between different payers, depending on the local circumstances which led to an agreement . This means that some list prices may substantially inflate the true cost of the service provided, while others may not cover the real underlying production cost of providing that service. One should be careful, however, to compare prices in different geographic areas for the same good corrected for Purchasing Power Parity, which is a measure intended to correct for differences in real wealth expressed by the purchasing cost of a fixed number of goods (i.e., a basket) in the different geographic areas. Therefore, it makes no sense to consider as similar the price of a 10$ item say in Switzerland and in Thailand, because of the huge income disparities between these two countries. A 10$ expense is a much greater strain on an average income in Thailand than in Switzerland. This is why most holiday travelers come from rich countries and go to low-income countries and not the reverse. Second, even market prices may not reflect the true opportunity cost because of imperfections in the health care market. For example drug acquisition prices by a hospital may differ from the official company price owing to discounts negotiated between the hospital pharmacy and the drug company.

When to adjust market prices in practice is not always clear. As a general rule, market prices should be adjusted if leaving prices unadjusted would introduce a substantial bias in the analysis and there is a clear and objective way of making the adjustment (Drummond et al, p56). Even in the case of severe potential bias, if the results of the analysis are not sensitive to the unit price of the resource then one might not bother to undertake a complicated price adjustment exercise.

Defining the Medical Intervention

The first step in any analysis is to define precisely the medical intervention being studied, its current alternatives, whether surgical or not, and the target population which would benefit from it. In principle, all relevant alternatives should be included. This means including less effective but also less costly ones, or, on the contrary, more experimental but more costly therapies. One should also include therapies in current use, even in the knowledge that these might not be clinically optimal.

The target population should be clearly defined in terms of pathology, stage of disease, gender and age.

Defining the Study Viewpoint or Perspective

This is a crucial step in the cost analysis process, as it will impact on the types of costs to be included and the way they will be measured. This is because what may be a cost to one agency or group may not be a cost from someone else's viewpoint. An easy example are travel costs for the patient and time spent in visits by his relatives which are borne by himself or his family but are not considered a cost from the point of view either of the hospital (the provider) or the MOH.

Although one generally distinguishes between several broad viewpoints that are mutually exclusive, a given study may consider costs from different viewpoints. This has become increasingly important as there has been a trend in health care systems in most developed countries towards shifting costs, at least partially, back to patients.

Different possible viewpoints include:

1. The Ministry of Health, or at a more regional level the Regional Health Authority.
2. The health care provider(s). For surgery this would generally be the hospital, although it might include other health care providers such as kinesitherapists for rehabilitation and general practitioners for follow-up.
3. The patients and their relatives.
4. The insurer or third-party payer.
5. The employer, for example when the employer pays part of the insurance claims or when one is interested on the absenteeism related to illness from the employer's perspective.
6. Other public or private agencies, for example those responsible for follow-up after discharge or for prevention.
7. Society in general, which is the broadest, but also the most difficult perspective to assess.

When planning a cost analysis it is therefore important to identify carefully the different perspectives that will be included and to identify (preferably in a matrix format) the relationship between the agencies involved and the types of costs to be estimated (for an example see Gold et al, p 187).

Often a regulator or social planner will only consider budgeting costs by considering the minimization of total (medical) costs or even specific medical resources outlays such as drug expenses. In this case, an optimal solution from the planner's point of view may differ from a societal optimal solution, as it does not include costs (and benefits) that are borne by other agencies or groups. This phenomenon is often referred to as the "silo effect".

Defining the Time Period of Cost Coverage

The time period during which costs related to the intervention should be tracked depends on a number of factors. The general principle is that the follow-up period chosen should not bias the analysis in favor of one of the interventions being compared.

In some cases this means that one should calculate costs over a very long period, possibly over a lifetime, although the further costs are incurred in the future, the less they will impact because of the discounting to present net value (see below).

In general the factors that will influence the time period for tracking costs are:

1. The occurrence over time of clinical side effects, adverse events or relapses related to the clinical intervention being studied. It is therefore important to know the natural history of the disease and the long-term consequences of the intervention. For example a study of the cost of CABG versus

PCTA would need a very long-term of many years of follow-up in order to take account of all the rehospitalisations following restenosis. Often, a cut-off point is agreed upon, because over time some effects become more rare and omitting a few of them would not significantly alter the results. In practice, however, one sees that most clinical trials are conducted over too short a time period to cover long term effects, so that there will be a need to complement these data by either retrospective analyses or statistical/ epidemiological modeling techniques such as parametric survival analysis.

2. The time frame of the relevant audience of the study may also limit the duration of cost tracking. For example, if an agency needs a budget impact analysis for a five year forecast, then it does not make much sense to expand the analysis at great effort over a (theoretically sounder) time period. In the same sense, a study conducted from a hospital's perspective might only consider costs up to discharge, although readmission costs may also be important in some cases.
3. Finally, the availability of epidemiological and clinical data, especially on the occurrence of long term effects may often not be available, and therefore preclude any long term assessment.

Another problem linked to the time period of cost tracking is that in any analysis we assume a constant medical technology with generally no quality-adjusted price indexation for medical goods, such as apparatus or drugs. This means that we consider, say, a PCTA in 10 or 20 years of time will be performed with the same technology at the same (discounted) cost as a current PCTA, while it has been shown that at least for some medical technologies relative prices have decreased dramatically over time due to advances in technology. This is, for example, the case for a large number of laboratory tests as they have become automated and more widely used, or for medical imaging procedures.

Defining Types of Costs

Costs are usually divided into three broad categories in the economic evaluation: direct, indirect and intangibles. Direct costs are resource costs that are attributable to the intervention under study, including side effects or other current and future consequences. Indirect costs are used to refer to the productivity losses or other losses related to illness and death (Gold et al, p 179). Indirect costs in economic evaluations should not be confused with the indirect cost terminology used in accounting which refers to overhead costs of the production of some good or service. Intangible costs are costs for which (at least currently) no economic value can be ascertained. These would include costs such as pain, anxiety and grief. These are generally included through quality-of-life or utility measurements.

Direct costs can be further broken down into two subcategories, direct health care costs and direct nonhealth care costs. Direct health care costs include the cost of hospital and community care, including testing, drugs, supplies, caregiver and support personnel and medical facilities, maintenance and purchase of equipment, cost of follow-up, treatments of side-effects, maintenance therapy and costs of rehospitalisation and retreatment (for example open surgery after failure of PCTA).

These also include, from the patient's perspective, any direct out-of-pocket expenses which are related to their treatment that are not reimbursed, including items such as drugs and appliances.

Direct nonhealth care costs include all other nonhealth care resources that are consumed as part of the intervention or its follow-up (Gold et al, p179). These cover such items as patient and family transport costs, informal caregiver's time (also known as home production costs) and home care support. The cost of resources consumed from other sectors of the economy such as social workers, police and law should also be included. Personnel costs, for example in the case of treatments for mental illness, are also usually considered.

Whether the patients' own treatment time should be included as a (direct) cost has been subject to debate in the economic literature, as two different interventions for the same diagnosis might have a different time span of care. If we view time as a (limited) resource to the patient and to society, then we should include the value of the time consumed by the intervention in our cost calculations, including waiting time (which might be important for some procedures or in some countries), travel time, and the time of the intervention and care itself. Generally, a patients' time is considered in an analysis carried out from a societal perspective, and therefore included by many authors as an indirect productivity loss (for a dissenting view see Gold et al, Chapter 4). We would recommend that whenever patients' time is valued, to keep it separate from the other costs and to include it as a separate item as part of the loss of productivity costing estimates (corrected for double-counting; see also Dranove, p74-75).

Indirect Costs

Indirect costs pertain to the loss of productivity to society associated with the loss or reduced ability to work or engage in normal life activities or leisure due to the illness (i.e., morbidity losses) or to premature death (i.e., mortality losses). Morbidity costs also include the time periods needed for recuperation and convalescence. Alternatively, these could also be integrated as a reduction in quality of life in cost-utility analysis or as a reduced benefit in cost-benefit analysis. One should however take care to avoid double-counting in those cases. When conducting pure cost minimization analysis however, all these costs should be included.

There are several challenges in measuring productivity losses resulting from disease or health interventions. The first challenge is related to the measurement of the productivity losses themselves. When an employee or any worker is absent from work this is a relatively obvious measure, but when productivity is reduced such as in migraine patients or in depression, then accurate measurement becomes more difficult.

The second challenge is related to the valuation itself of time lost to illness. Usually one uses gross earnings (including benefits and employment overheads) to value time lost from work, but in the case of nonsalaried workers (for example self-employed workers) or nonemployed patients (for example, housewife, retired, unemployed, students), the value of lost time is difficult to assess. The same can be said about the valuation of unpaid informal care (Brouwer et al, 1999). One of the solutions is to find similar marketed services and to value time according the market price of those services (for example childcare, housekeeping).

The third concern is what the true cost to society is when an individual is taken from the workforce, when he can be replaced by colleagues or do some catching-up when returning to work after a short-term absence, or for long-term or definitive absence when he can be replaced by some replacement worker. Therefore the amount of productivity lost depends on the cost of organizing the replacement and the loss in productivity over some training and adjustment time when hiring a new worker. This has been called the "friction cost method" which was proposed by Koopmanschap and colleagues (1995). It should be mentioned that this method gives cost estimates that are much lower than those obtained from the traditional wage imputation.

For nonemployed patients the impact of illness will more likely involve a decrease in the ability to perform activities of daily life (ADL) and leisure. In general, one tries to find a market substitute for such activities in order to value these, or in the case of lost leisure, for example, one could use willingness-to-pay (WTP) or willingness-to-avoid survey methods to put a price tag on such time.

Transfer Costs

Transfer costs are income transfers, involving the redistribution of money between different groups in society or between different payers but this does not involve a consumption of resources. For example, disability payments involve the transfer of money (from workers and employers) though the government to the group of disabled, but this does not change the aggregate amount available to society as a whole. When the viewpoint is more restricted, however, to, say, a single agency or part of society, then transfer payments may be considered as gains or costs from the viewpoint of that agency. In any case, the administrative process of transferring that money does involve "transaction" costs, which ideally should be incorporated in the analysis, although the transfer payments should not.

Unrelated Future Costs

In cases where an intervention is life-saving, those individuals will incur "unrelated" health care costs in the future. For example, a woman having had a hysterectomy for local carcinoma at 55 years with no relapse will have a life expectancy equal to that of the general population and will also incur health care expenses over her remaining life-time similar to that population.

Should these costs be included in the total cost of the intervention?

This is particularly problematic when studying prevention programs compared to a "do nothing" or a therapeutic intervention program, as this would unfairly disadvantage all prevention programs. Furthermore, there might be cases where we face competing risks, where a reduction of the incidence of one disease increases the occurrence of other diseases. This has been the case historically with the decrease of deaths due to infections and the rise in cardiovascular and cancer deaths.

Another argument for leaving out these future costs is that they generally occur late in life, and their impact will be reduced via discounting. Therefore, the current practice is to leave out future unrelated costs but, if desired, to include a sensitivity analysis whereby discounted average annual age-related per capita health expenditures are included. In this way, one can see whether or not future costs radically change the results of the more partial analysis.

Defining the Resource Items to Be Collected

Once the above parameters have been established, a cost inventory is developed. This is a comprehensive list of all the resources required to carry out an intervention. Typically, a cost inventory includes:

1. Personnel costs: direct caregivers, support staff, administrative staff, and volunteers' time.
2. Laboratory costs.
3. Drug costs and drug administration costs.
4. Supplies and materials.
5. Equipment used.
6. Maintenance costs for facilities and equipment.
7. Facilities, including rent and utilities.
8. Transportation costs.

Other costs of providing the services include computer equipment, courier services, archiving of materials and results, uniforms, insurance premiums, educational and training costs, and R&D costs.

In general, one attempts to create a list which is as comprehensive as possible, as one does not know beforehand which costs will play a large role in the care process. Also, a small cost shared by a large number of patients may amount to a large sum, while on the other hand, an expensive but rare event (e.g., an adverse event), although clinically important, might not add much to the overall cost of the intervention.

Collecting Data on Resources Used

10

Once a complete cost inventory has been established, when comparing two different interventions it is useful to identify which costs are similar to both, so that these can be factored out from any comparative analyses.

For example, if the administrative tasks are identical among interventions being compared, these can be dropped from the analysis, but not the drug costs, if the drugs used are different or are be used in different amounts. Often in health care settings such as hospitals, accounting procedures exist to allocate resources to cost centers, which can then be linked to medical departments. In some cases, accounting procedures are sufficiently developed to collect costs on a case or disease classification basis such as per DRG (diagnosis related group) or ICD class.[2] However even these are often too coarse when assessing a specific intervention in a well defined clinical condition, so that a local resource consumption analysis has to be conducted by micro costing (see below).

Drug resources are usually calculated from either patient charts or automated pharmacy delivery systems. For operating room use, time of operation is usually recorded as well as all materials and supplies used. Some hospitals have also developed detailed standard cost per hour estimates for operating room use. More difficult to assess are resources used in diagnosis and surveillance, such as medical imaging, radiology, and laboratory tests. These often have to be collected from patient charts and files if no comprehensive tracking and/or billing systems already exist. In some

[2] For an example in Canada see the Case Costing Project in Ontario.

sense, the best in-hospital systems are those that are still based on detailed fee-for-service systems, so that every medical resource used is coded and billed. Often, as in Canada, only some of the resources used are collected routinely and billed on a fee-per-service basis, while large areas of direct care are included in the overall hospital budget allocation. For ambulatory care there exists a large variety of tracking systems, from the possibility of integrating ambulatory and hospital data on an individual patients basis such as in Saskatchewan or in Health Maintenance Organizations in the U.S., to no tracking system at all. Most Provinces or organizations however fall somewhere in between, for example in Quebec, it is possible to track drug deliveries and GP visits for patients over 65 years.

Generally, a mix of different approaches is to be used, depending on the local situation and data availability, especially when one wants to integrate inpatient and ambulatory outpatient costs.

Estimating Unit Costs of Resources

Direct resource costs can be divided into variable, semi-variable and fixed costs. Variable costs are those that are directly related to the individual patient treated or seen. For example, laboratory tests, diagnostic tests and supplies are considered as variable, because if no patient is treated then these costs are not incurred. Provider and health care workers time is either considered as a variable cost on the same grounds or may be semi-variable, depending on the perspective of the study and the time horizon. For example, a physician's time, if paid on a fee-for-service basis, would be considered as a variable cost whatever the perspective, but if the same physician is salaried in a hospital and one conducts the study from the hospital's perspective, then one might argue that his time should be considered as a semi-fixed cost, i.e., whether or not he treats patients in the short term his salary will still have to be paid.

As an example, let us assume a ward with 20 beds is attended by 5 full-time salaried nurses. Suppose now that a new medical technology, such as laparoscopy is available. This technology reduces the length of inpatient stay by some proportion, so that, for the same number of patients treated, only 16 beds and 4 nurses are needed. In the short term, the salary of the excess nurse would probably still be supported by that ward; after some time, however, that nurse would probably be transferred to another ward (as would the freed beds). We will assume that there is no waiting list for similar patients or for patients with another pathology that would be treated in the same ward.

Fixed costs include the cost of buildings, facilities (i.e., construction costs) and of purchased equipment. Maintenance costs could be considered semi-fixed when a long-term maintenance contract has been signed with the seller or a third-party or could be considered variable when maintenance is carried out in-house and is therefore a direct function of the workload. In the long run all costs can be considered variable, as personnel can be dismissed, wards closed and equipment resold.

When starting a cost study it is therefore recommended to spend some time on defining the resources used and to classify these according to their variable, semi-fixed or fixed nature.

How costs are classified has an impact on the calculation of the average cost per patient as can be seen from Figure 10.1.

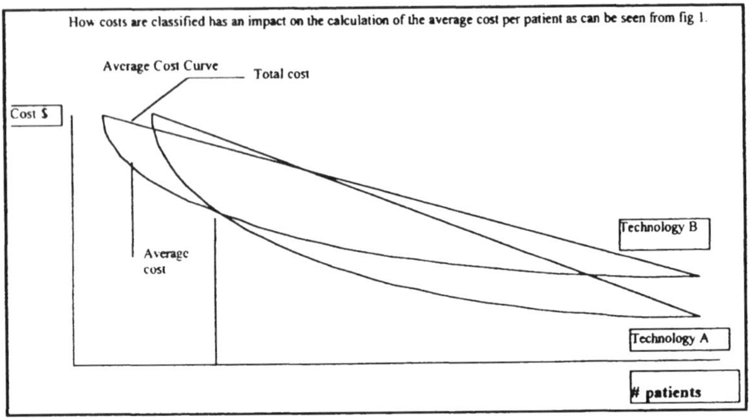

Fig. 10.1 Average cost curve

10 We see that the average cost per patient (AVC) in case A drops quickly below the average cost per patient in case B (the straight line shows the trend), because higher fixed costs in A become spread over more patients as the number of patients increases. In case of A, when only a few patients are treated, the average cost per patient will be very high. Given two different medical technologies we can then compare the average cost per patient, for a given number of patients to treat. The exact point (of volume of patients) where the average cost will be equal between both technologies depends on the proportion of fixed and variable costs in each of these. Therefore in cases of expensive equipment these should be used at their maximum capacity as far as possible, especially when the variable cost of performing the examination is relatively low.

Total, Average and Marginal Cost

From the previous discussion we can easily see that the total cost of a medical intervention will vary differently for variable than for fixed costs when the volume of treated patients changes. This is expressed in a cost function, which expresses the total cost of a single intervention or technology in function of quantity. The total cost is the sum of the variable and (semi)fixed costs (see Fig. 10.2).

If we take the example of a mammography examination then the fixed costs include the cost of the examination room and the purchase (annualized) of the equipment, which is purchased as a discrete item.

In the first case we have $2 of fixed equipment, $5 of semi-fixed nurse salary and $1 fee per screening for the physician, in the second case all costs are the same except that the nurse is now paid also on a fee-for-service basis at a rate of $2 per screening. Total costs are represented in Figure 10.2.

As we can see from Table 10.1, the total cost increases more rapidly in case B which has the highest level of variable costs, starting from infinity in both cases and decreasing to a limit of $1 in case A and to $3 in case B.

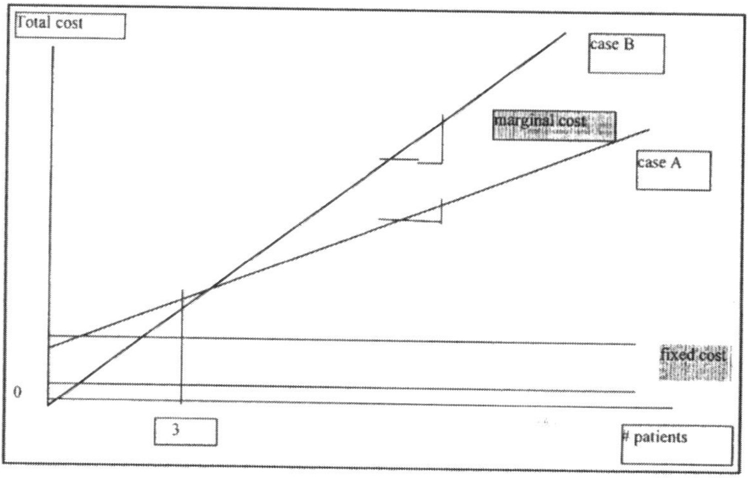

Fig. 10.2. Total cost function.

If we expect to treat 5 patients, then the difference between the average cost of case B and A is equal to:

(17/5)-(12/5) = 3.4-2.4 = $1.

This is known as the average incremental cost of Technology B versus Technology A (often noted as DC), and the total incremental cost is equal to 5 x $1 = $5.

It is important to bear in mind that we compare two medical strategies or technologies, we assume treating the same number of patients in both cases.

What happens when we want to treat one more patient? Then we need to calculate the marginal cost for that patient treated with a given technology i.e., the cost of one extra patient from the total cost curve (Table 10.2).

The marginal cost concept is important when calculating costs when the variable cost component is changing over time, such as in length of stay (LOS) in hospitals. This is because reducing the LOS often relates only to the fixed "hotel costs" of the stay on the last days of hospitalizations, whereas most investigations and treatments (the variable costs) happen on the first few days. Therefore, using an average cost estimate per day overestimates the true resource cost of the days saved. Also, after discharge, the first periods of follow-up are more resource intensive than afterwards, so that the average cost of different periods after discharge shows the same behavior over time as seen in Figure 10.3.

In general, average costs show a continuous decrease only up to some point, after which it becomes necessary to purchase additional equipment as the existing equipment is used at full capacity, or to increase the size of the facilities. In this case the average and marginal cost curves show a stepwise behavior i.e., they become jumpy at the point where the capacity has to be increased and show a high rise. This has to be taken into account when comparing medical technologies with different degrees of "lumpiness".

Table 10.1. Example of mammography screening

	Case A					Case B				
# patients	equip-ment	nurse	physi-cian	Total cost	avg. cost	equip-ment	nurse	physi-cian	Total cost	avg. cost
0	2	5	0	7	∞	2	0	0	2	∞
1	2	5	1	8	8	2	2	1	5	5
5	2	5	5	12	2.4	2	10	5	17	3.4
6	2	5	6	13	2.16	2	12	6	20	3.33
10	2	5	10	17	1.7	2	20	10	32	3.2

Table 10.2. Marginal costs of two strategies

Number of patients	Total cost Case A	Total cost Case B
5	12	17
6	13	20
marginal cost	1	3

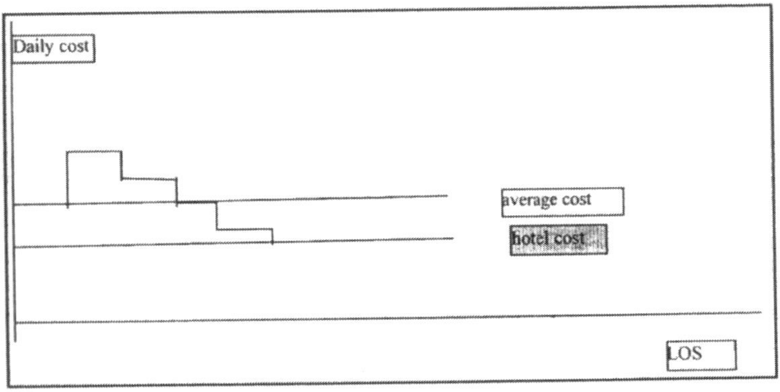

Fig. 10.3. Average and marginal cost of LOS.

Correcting for Time Preference

When costs or benefits occur into the future, it is necessary to incorporate the concept of time preference as distinct from inflation. This means that we have to convert future costs (and benefits) to their present value. This is the aim of discounting and the measure of time preference is called the discount rate. This is because the cost of the intervention is incurred in the present but the (avoided) costs will be only incurred some time in the future. However, society (and individuals in general) expresses a preference for benefits received today rather than in the future. Also, in order to compare different interventions with different time horizons, we need to know their present value.

There is a large economic literature on how to select and choose the best discount rate (for references see Lipscomb et al, chap 7, in Gold, 1996). The lower the discount rate the higher we value future benefits. If the discount rate is set to zero, then we make no difference between the present and the future. On the other hand, a 10% discount rate means that we are indifferent between incurring a cost of $100 today versus $110 next year, so that $110 next year is equal to $100 today (remember this is true even when we assume zero inflation).

Assuming that the costs occur at the beginning of the year (or time interval) we then calculate the present value of the stream of future costs (or benefits) a where r = the discount rate and n= the number of time periods

$$PV = \sum_{n=1}^{N} C_n (1 - r)^{-n}$$

For example, for an intervention with costs spread over three years (e.g., follow-up costs)

Year 0 =current year	
Year 1	100$
Year 2	30$
Year 3	20$

with r=0.05 or 5%

we have PV $= 100/(1+r) + 30/(1+r^2) + 20(1+r^3)$

$= 100/1.05 + 30/1.10 + 20/1.16$

$= 139.73.$

The factor $(1+r)^{-n}$ is the discount factor, which expresses time preference. This is the simplest case of time preference known as constant discrete-time discounting. In general, published analyses and current consensus recommend the use of a 3-5% discount rate, and a sensitivity analysis with undiscounted costs.

Health effects occurring into the future should also be discounted. Think about gaining a life-year now or in 30 years from now, are these equivalent ? There is, however, much debate on whether one should then use the same rate for health effects as for costs, whether that rate is constant over time, whether we should allow for differential rates according to the age-gender characteristics of the patients, and how these practices translate into the choices (i.e., the cost-effectiveness or cost-utility ratios) we estimate (for a recent discussion see Van Hout, 1998).

Correcting for Inflation

When costs are collected retrospectively or when one uses past costs, these should be adjusted for inflation to the base year where we conduct the study. Suppose we have a cost estimate of some medical intervention in 1995 but not for 2000, and we wish to adjust it for the current year. To this we have to deflate (or rather inflate) 1995 prices for the increase in the cost-of-living and general price evolution in those 5 years. To do this we could either use a general Consumer Price Index (CPI) or a more specialized medical price index, if available. In some countries national statistical institutes calculate more detailed medical services and hospital price indexes. The current adjusted price is then the past price multiplied by the ratio of the value of the price index in the current year and the value of the price index in the past year:

$$P_{2000} = P_{1995} \cdot [I_{2000}/I_{1995}]$$

Of course, the more specific the price index we use, the better the cost adjustment will be. For future costs there is no need to adjust for inflation if we make the assumption that the different cost items follow a path of balanced inflation. All calculations are then conducted in real base-year terms.

If we have good reasons to suspect an imbalanced inflation rate between two cost components, however, then in principle, each cost component should be allowed to inflate at its own rate over time, before being discounted for time preference.

A particularly difficult problem is to adjust future prices for quality improvements as medical technologies and drugs improve. Although the nominal price might increase over time, the real quality-adjusted price of the same item might well decrease over time. In practice, most studies assume a balanced, uncorrected for quality, inflation process (for a discussion and examples see Tripplet J, ed, 1999).

Micro-Costing versus Gross Costing

When costing resource use of medical interventions, one can choose along a spectrum from general cost measures to very specific ones, ranging from the gross cost of an event such as a hospitalization episode for gastrectomy based on national average estimates, down to the detailed micro-costing approach using time-motion observational studies of all interventional activities. One has to trade off specificity and accuracy versus cost and effort of collecting the data, along with the existing data availability.

The different levels of precision of cost estimates for an inpatient are summarized in Table 10.3.

For ambulatory care, the same tracking methods can be used to derive unit costs of services provided, for example the Resource-Based Relative Scale Value Scale (RBRVS) linked to the CPT-4 (Current Procedural Terminology) classification in the U.S. Generally, this defines reimbursement rates, but one needs to convert these to costs, using cost-to-charges ratios such as those published by the AHCPR (Agency for Health Care Policy and Research) for acute hospitalizations, but in many other countries these may not be available.

Conclusions

In summary, a wide variety of costing methods exist. The choice of which to use depends on the availability of local cost data and accounting and cost tracking systems in use. The major task facing the researcher is then to decide at which level of detail and accuracy costs should be tracked and how to integrate the different cost components of a broad range of medical and professional services, especially when, as in many cases, one goes beyond the acute inpatient care episode to include follow-up ambulatory care.

Selected Readings

1. Baker JJ. Activity-Based Costing and Activity-Based Management for Health Care.Gaithersburg, Maryland: Aspen Publishers Inc. 1998. Gaithersburg, Maryland: Aspen Publishers Inc. 1998.
2. CCOHTA. A Guidance Document for the Costing Process, version 1.0, CCOHTA, Ottawa, August 1996.

Table 10.3. Levels of cost Estimates

- Average per diem for the whole hospital
- Average departmental cost per diem e.g., all patients in Intensive Care Unit or in pediatric ward
- Diagnosis Related Group (DRG) or Case Mix Group (CMG)cost par episode. This gives an average cost per episode based on a disease classification. Many countries are developing such a system for hospital admissions, often with a local disease classification scheme based upon the ICD-9 or ICD-10 classifications such as the PMSI in France or the CMG in Ontario in Canada.
- Activity Based Costing This breaks down the different components of health care delivery into basic homogeneous units or procedures, e.g., a knee arthroscopy, and allocates general costs to each basic component of the activity either through a bottom-up or top-down approach (Baker J.J., 1998).

3. Drummond M, O'Brien B, Stoddart G et al. Methods for the Economic Evaluation of Health Care Programmes, Oxford Medical Publications, 2nd edition, Oxford, 1997.
4. Gold MR, Siegel JE, Russell LB et al. Cost-Effectiveness in Health and Medicine, Oxford University Press, New-York, 1996.
5. Haddix AC, Teutsch SM, Shaffer PA et al. Prevention Effectiveness, Oxford University Press, 1996.
6. Johannesson M. Theory and Methods of Eonomic Evaluation of Health Care, Kluwer Academic Publishers, Boston, 1996.
7. Sloan F, ed. Valuing Health Care, Cambridge University Press, 1995.
8. Triplett JE ed. Measuring the Prices of Medical Treatments, Brookings Institution Press, Washington D.C., 1999.

Articles

1. Brouwer W, van Exel NJ, Koopmanschap MA et al. The valuation of informal care in economic appraisal. Int J Tech Assoc Health Care 1999; 15:1;147-160.
2. Earle C, Coyle D, Smith A et al. The cost of radiotherapy at an Ontario regional cancer centre: A re-evaluation. Critical Reviews In Oncology/Hematology 1999; 32:87-93.
3. Helms LJ, Melnikow J. Determining costs of health care services for cost-effectiveness analyses. Medical Care 37(7):652-661.
4. Van Hout BA. Discounting costs and effects. A reconsideration 1. Health Economics 1998; 7(7):581-594.

10

CHAPTER 11

Databases and Registries

J. Sampalis

Overview

The advances in computer technology in the recent years have provided the tools by which large databases or registries could be used in medical research. Tumor registries and registries of rare diseases are the classical examples of large databases that have been used to assemble and follow large cohorts of patients over long time periods. In addition to providing epidemiological information on specific conditions, registries also provided information on patient outcomes and in some cases on interventions or treatments used. Registries provide the advantage of sufficient sample size and duration of follow-up that allow statistically valid estimations of disease burden or of the association of patient outcomes with patient characteristics or with medical interventions. Collection of data on similar patients from different centers or geographical regions, or from different time periods into a single registry could enhance the external validity or generalizability of the data. The lack of standardization in the data collection process both with respect to the information obtained and the patients included may compromise the validity and utility of registries. Nevertheless, well designed and organized registries could be valuable tools in the study of disease and the evaluation of health care interventions.

Although local hospital registries may be common in most surgical departments and clinics, large multi-center registries have had limited application in surgical research. The exceptions are tumor registries which may contain information on patients that have received surgical treatment and trauma registries. Trauma registries are an excellent paradigm of large multi-center databases which have been used not only in the epidemiological description of injuries but also in the evaluation of trauma care. Results of studies based on trauma registry data have had a significant and direct impact on the evolution of trauma care.

The purpose of this Chapter is to describe the essential requirements and steps in developing a disease registry.

1. Defining the Purpose

Disease registries have been implemented in response to the need for continuous improvement of the health care provided to specific patient populations. Continuous quality improvement requires that data on the treatment and outcome of patients are collected in sufficient detail and continuity in order to allow the evaluation of the impact of interventions.[1] This evaluation should take into account the epidemiological profile of the disease, the interventions used against the disease, the

Surgical Arithmetic: Epidemiological, Statistical and Outcome-Based Approach to Surgical Practice, by Lawrence Rosenberg, Lawrence Joseph, and Alan Barkun. ©2000 Landes Bioscience.

outcomes produced as well as the resources utilized. Registries have been used to synthesize and analyze this information through the collection of data from individual patients. The results of these analyses are often used to modify the interventions in a way that the patient outcomes will improve while expenditures are minimized. It should therefore be always maintained and understood that the ultimate purpose and goal of any registry is to provide the material for continuous quality improvement in the target patient population with implications within the individual hospital, the system, geographical region or other boundary from where the patients are identified.[2-7]

Although quality assurance is the primary purpose of any registry, the information in these databases is of value to administrations at various levels including those of hospitals. The data required for administrative purposes are quite often a subset of those obtained for quality assurance. However, the reporting requirements both with respect to frequency and structure are often quite different. Accommodation of the administrative needs enhances the utility of the registry and facilitates participation and compliance of the participating institutions.

It follows from the above discussion that registries should have a dual purpose. The first purpose is to conduct a quality assurance evaluation. The second purpose is to provide information to administrative bodies regarding the patient population studied. As an example, the purpose of a provincial or statewide trauma-registry could be described as follows:

1. To conduct an evaluation of the quality of care provided to patients with major trauma.
2. To describe the incidence of major trauma in the region.
3. To describe the resource utilization for the care of trauma patients.
4. To describe the distribution of trauma patients within the regional hospitals.

This set of goals could address the needs for quality assurance while providing regional health administrators with important information that could be used for decisions on resource allocation. It should be noted that although each one of these items may be classified as either predominately evaluative or predominantly administrative, they all address both needs of the Registry. As an example quality assurance information from the entire system or individual hospitals may be used for hospital designation whereas the distribution of patients in the hospitals of the area may provide clues for the explanation in the variation in patient outcomes. The incidence of major trauma and resource utilization provide information that could be equally important to clinicians, epidemiologists and administrators.

2. Identifying the Patient Population

Identification of the specific population of patients that will be included in a registry appears on the surface to be straightforward. However, in many cases the patient selection criteria should be defined after careful consideration of the epidemiology, the patient profile and the outcomes of the disease under study. Just like any other project, a registry is limited by the availability of funding and resources. It is therefore essential that the patients included in the registry be selected such that all required information is obtained while redundant information is minimized. In order to accomplish this, the patient selection criteria should take into consideration the epidemiologic or clinical profile of the disease as well as the association between patient or disease characteristics and expected outcomes.

An example could be drawn from trauma registries. The epidemiology of injuries shows that 80-85% of all patients treated in a hospital emergency room for injuries have minor trauma with no risk of significant disability or mortality. Approximately 15-20% of all trauma patients are admitted into the hospital after the Emergency Room and of these, almost 40% have minor injures and are released within three days after admission. The important subset of patients are those with major injuries which comprise approximately 10% of the total. This patient subset is most affected by the quality of care and exhibits the widest variation in outcomes. These patients also utilize over 80% of the total health care resources expended for the management of trauma patients. By selecting 100% of all major trauma patients (or 100% of all admitted patients) and 5-10% of all minor trauma patients for inclusion in trauma registry, the expenditures are minimized, the entire spectrum of trauma severity is represented and all of the patients providing the most valuable information for quality assurance are included. The database created in this manner will provide the necessary information to address all the goals of the registry while minimizing costs. Similar selection processes could be applied for the majority of the diseases for which a registry could be established. Care should be taken to ensure that important subgroups of patients are well represented and that the expected sample sizes are sufficient for valid and precise estimation of the required parameters and associations.

3. Determining the Variable Set

By far the most important step in designing and implementing a Registry is the identification of the data items that will be collected. The final set of data variables should be based on a balance between the need to obtain all the necessary information and minimization of costs. Registries with extremely large numbers of variables are highly labor intensive and are prone to low compliance, high proportion of missing data points, and ultimately fail due to lack of participation. On the other hand registries with very few or insufficient data elements may be easy to implement with a high rate of compliance. However, the lack of sufficient data limits the usefulness of these registries. The following criteria should be applied in selecting variables for a Registry database.

1. The variable is absolutely essential for the goals of the Registry.
   i.e., If this information is not available, the Registry will fail in achieving its goals.
2. The data required are readily available in the source and do not involve interpretation or extensive research.
   i.e., The data should be well recorded in the patient records. The abstractor should not be expected to interpret other information or to search several sources.
3. The information has an acceptable degree of validity.
   i.e., The data must be as valid as possible with a minimal possibility of wrong information. Verification of questionable or ambiguous data should be possible.

4. The information should be available for the majority of the patients.
   i.e., Variables with a high proportion of missing values are of limited use and are often costly to obtain. Inclusion of variables with missing information for over 40-50% of the cases should be carefully considered.
5. Obtaining the information should be cost-efficient.
   i.e., Variables that require extensive review or search of the sources but provide information which is of marginal utility should be excluded. The time and effort required to obtain the information should be justified by its potential value towards realization of the registry's goals.

The variables included in the database should be classified using a two dimensional grid. One axis of the grid describes the content of the variable whereas the other axis describes the timing of the data collection. With respect to content, the following general sets of variables should be included in the registry:

1. Patient characteristics.
2. Disease descriptors.
3. Intervention descriptors.
4. Outcomes

Patient characteristics should include at the minimum the usual demographic information and when available any information on concurrent disease and concomitant medications. Disease descriptors should include details on the duration of the disease, an indicator of severity, and when appropriate any information on the sub-type of disease or symptoms. The information on disease severity should be sufficient to classify patients according to the severity of the disease and should be sensitive enough to allow the evaluation of treatment effect. Intervention descriptors should not be overwhelmingly detailed but should provide sufficient information to classify patients according to the treatments received in homogenous groups. The timing of the interventions should be available in sufficient detail to allow assessment of temporal associations between treatment and outcomes. In addition, information on diagnostic procedure, physician visits, and hospitalizations should be obtained when possible. These data are often essential for quality of care evaluation and cost measurement. Variables describing patient outcomes that are included in registry databases are usually limited to those capturing the change in disease severity or state. More subjective outcomes such as quality of life are usually not obtained because of the high cost associated with administering and recording the required questionnaires. However, measures of functional status or other indicators of well being that are collected routinely as part of the patient's evaluation should be included in the registry.

With respect to the timing of data collection, the variables are classified into two broad categories. The first category comprises the variables which are collected only once and are not expected to change during the follow-up period. Such variables are often patient characteristics and baseline disease descriptors. The second category of variables are those that are collected at many points in time during the follow-up. The frequency with which these variables are collected should ensure that important changes are captured and are accurately described. Such variables include disease state, treatment description and patient outcome.

Once the set of variables has been identified, pretesting is recommended in order to ensure that the data are available and easily obtainable. The validity of the information should also be verified during this pretesting period. Quite often the dataset will be modified after the pretest with more exclusions and a small number of added variables. The pretesting exercise may involve considerable effort, however, the benefits are substantial. Prior to finalizing the variable set a panel of experts may be consulted in order to ensure that important data elements have not been omitted.

4. Direction and Follow-Up

Whether the patients included in the registry are identified prospectively or retrospectively depends on a number of factors, including the nature of the disease studied, the incidence of the disease, the specific goals of the registry and the available funding. As with any other observational study, and registries are observational epidemiological studies, prospective studies are considered methodologically superior to retrospective studies. Disadvantages of prospective studies include high costs and long duration of accrual, which often inhibit their implementation. Quite often hybrid retrospective-prospective studies are implemented. Ideally all patients included in a registry should enter the cohort at the same landmark defined by the disease under study. Patients included in trauma registries enter the cohort at the time of injury, whereas cancer patients are entered in tumor registries at the time of diagnosis. Registries of surgical patients may define the point of entry as the time when the patient becomes a surgical candidate. In other registries the patients may be entered at the time of surgery.

The duration of follow-up should be sufficient to allow the detection of any significant changes in patient outcome. Care should taken not to extend the follow-up beyond the time period when meaningful changes in outcome should occur. As an example, trauma registries are usually concerned with the acute phase of trauma care and limit the patient follow-up to discharge from the acute care hospital. On the other hand registries of patients undergoing cardiac surgery should extend the follow-up several years beyond the patient's discharge from the hospital. Registries with insufficient follow-up duration are of limited use for quality assurance although they may provide useful epidemiological data.

5. The Data Collection Process

Once the final set of variables has been defined and tested, the process of actual data collection has to be designed, tested and implemented. At this point in time it should be pointed out that data collection could be performed retrospectively or in real time (prospectively) regardless of the direction in which patients are entered in the registry cohort. While prospective or real time data collection is appealing, for many of the variables it will not be possible. The majority of the registries collect data from patient charts in a retrospective manner on patients that have been identified prospectively. A limited number of variables should be obtained in real time in order to ensure accurate tracking of all patients. As an example, patient identifiers, medical record numbers and date of entry into the cohort may be collected in real time. This will permit identification and tracking of the patient so that data collection could be completed at a later point in time.

The design and implementation of the data collection process should involve the following steps.

Establishment of Accurate and Timely Identification of Eligible Patients

Depending on the disease studied this may require the cooperation of clinicians, nurses, or medical records personnel. It is essential that the individuals responsible for identifying the patients are involved in the implementation of the registry from its initial stages.

Establishment of Access to Data Sources

Establishing access to all sources of data is essential and has to be finalized early on in the implementation of a registry. The cooperation of clinicians, clinic or medical record personnel is essential. The authorization to approach patients when necessary should also be obtained very early on. At this point in time the issue of confidentiality should be addressed. The participating members that will provide patient information usually will require a guarantee that confidentiality of information is maintained at all levels. The process by which data will be managed should ensure that information on individual patients, clinicians, or institutions could never be disclosed under any circumstances. Those running the registry should devise methods by which the participants could be receiving information on their own patients but only summary data on the patients from other participants. Data on individual patients should never be disclosed.

The approval of local ethics committees or government agencies should be obtained prior to accessing patient data. An essential requirement of any ethics board is that absolute confidentiality be maintained.

Designing the Data Collection Form

The data collection form is the main tool of a registry. A considerable amount of effort should be devoted to ensure that the data collection form is easy to use and that it contains all the necessary data elements. Simplicity of use is the most important attribute of a data collection form. This could be achieved by following some basic rules:

1. The sequence of information requested should resemble that of the data sources.
2. Clear, easy to read forms are better than crowded, cluttered ones.
3. Data should be collected in the same format as in the data source. Calculations or interpretations should be performed after the data have been collected and entered in a computer database.
4. Open-ended questions should be avoided as much as possible.
5. Multiple choice questions should be used instead of extensive look-up tables. The exception to this is international classification codes for diseases, interventions or medications.
6. All questions must be answered. Give the abstractor the possibility of entering an answer even if the data are not available. As an example "unknown" or "not available" should be possible entries.

7. For numeric variables, possible choices of units should be provided. The data abstractors should not perform any conversions.
8. The same sequence of information should be used and specified for all dates, e.g., dd/mm/yy throughout.
9. Logical nodes should be introduced which will direct the abstractor. As an example skipping over a series of nonapplicable questions.
10. Some indication of where that data may be found should be provided when possible. As an example, radiographic report, nurses notes, or discharge summary.

Standardization of the Data Collection Process

This could be established by the creation of a data collection manual and training of the data collectors. The data collection manual should describe in as much detail as possible the procedures and definitions employed for data collection. The manual should be used by data collectors both as a user's manual and a reference.

All data collectors should receive training at the initiation of the registry. The training should be repeated at regular intervals in order to ensure that a high level of expertise with the data collection process is maintained. The training sessions should provide the opportunity for data collectors to exchange comments based on their experience with the data collection process. These comments should be considered in future revisions of the registry. Training sessions are especially important for multicenter registries and regular monitoring of all centers?

6. Data Management

After data collection, data management is the second most important function in the implementation of a registry. Data management consists of data entry, data validation, data clean-up, and preparation for analysis. These functions are primarily performed with computers and involve the use of data management software. The selection of computer hardware and software that will ensure reliable, user-friendly and efficient data management in probably one of the most important decisions in the establishment of a registry.

The first step in selecting the computer equipment and software is deciding whether to use a commercially available software package or to undertake the task of developing a customized database program. There are advantages and disadvantages to both approaches. Commercial data management packages are purchased or licensed from software developers. Two types of commercial registry software are available. One type is designed for a specific disease or patient population. The second type is more general and allows some modifications to accommodate the needs of specific registries. The advantages of commercial packages are the following:

- pretested and established
- user support
- standardized use across several centers and similar registries

The disadvantages of commercially available packages are:

- high acquisition or license cost
- limited flexibility
- expensive maintenance
- dependence upon developer

During the last decade database software such as Dbase, Objectvision and Paradox[8-10] have provided platforms on which data management software could be developed by experienced programmers. More recently statistical packages such as SPSS[11] offer modules that allow the development of data management software. The advantages of custom-developed data management software are:

- specifically developed to address the needs of the registry
- completely flexible
- controlled cost

The disadvantages of custom developed software packages are:

- lack of standardization with other registries
- high initial investment
- long development period

Regardless of the choice of the data management software the following are the minimal requirements:

1. User-friendly interface.
2. Compatible with several operating systems.
3. Allow data entry and correction.
4. Built-in automatic error checking and data verification.
5. Export data for review on a case per case basis.
6. Backup of data schedule and process should be automatic and built in.
7. Allow export of data into most commonly used database and statistical software.
8. Perform basic analysis of data and produce reports.
9. Ensure confidentiality of patient or institution nominative information.

An additional asset of data management software that should be considered in view of the increased use of computer networks, in multi-user capability.

11

Data collection and data entry of most registries involves a two phase process. The first phase is data collection during which the data are abstracted from the data source and are transcribed onto a paper-data collection form. The second step is data entry during which the data are entered from the paper form into the computer. Collapsing of the data collection and data entry phases into one step by which the data are entered into the computer directly form the source is a relatively new approach. This method which bypasses the paper data collection form has been used for a small number of registries and in some clinical studies. The advocates of this method propose that it is more efficient and reduces the errors form transcription. Those not supporting this approach state that the time required for the one step approach is at least equal to, if not more, than the two step approach. In addition, the errors are not minimized and correction of wrong entries becomes more complicated and time consuming since review of the data source would be required. To this date the benefits of direct computer entry have not been demonstrated. The complexity of medical records and the limitations of computer software are factors which continue to prevent the wide implementation of the direct entry method.

In the two phase method data entry should be conducted by an experienced clerk and not by the source individual who collects the data. This allows independent error checking of the information in the paper form. The data entry clerks should be responsible for most suspects of data management including reporting of errors or discrepancies in the data collection form, conducting a validation of the data using the build-in software modules, ensuring that the database files are appropriately backed up, correcting the data, and producing reports using the build-in software modules. The data entry clerks should also be responsible for transmitting the data to the central registry database according to the established schedule. Data entry clerks should receive initial and follow-up training on the use of the software. In addition, similarly with data collection, a data entry or data management software manual should be produced. The manual should describe in extreme detail the operation of the software as well as the procedures and definitions used in data entry.

Multi-center registries should have a centralized control center where the main registry database is maintained. This centralized database should receive data from the satellite centers on a regular basis. The ideal situation is that where data are abstracted and entered at the local center. Initial verification, validation and error correction is performed locally and then the data are transmitted in an electronic format to the central database. This could be accomplished by diskette, modem transmission, or direct internet connection from one computer to another. The data transferred to the central database should not include any patient identifying information. All records, however, should be identified by a unique registry number that will allow accurate tracking when necessary.

Local single center registries and main databases should be located on computers separate from others used in the study for analysis, data verification or entry. Access should be limited to the individual responsible for maintaining the registry in the institution.

When necessary, further validation of the data may be conducted at the main registry. Potential errors should be reported back to the local centers or the data management personnel. The corrected records should then be re-transmitted and incorporated in the main database. All corrections or changes to the data in the main database should be logged or archived.

7. Record Linkage

Most registries require data on the same patient from multiple sources. The records from multiple sources have to be linked into one record in order to prepare the data for analyses. As an example, trauma registries require linkage of records from the prehospital services, the hospital and the rehabilitation center.[12-16] Other registries may require the linkage of records form more than one hospital or from various clinics or services within the same hospital. Record linkage requires unique patient identifiers that are common and identical in all data sources. As an example, the patient's name may be used as such a unique identifier. Quite often, however, a single unique identifier is not sufficient as it may create ambiguous matches. Certain names may be common in certain communities or cultures. Relying therefore only on patient names for record linkage may lead to mismatches. A minimum of two variables should be used when the probability of ambiguity exists. As an example, patient's names, date of birth, and gender may reduce the ambiguity and increase the chances of properly linking the records of the same patient.

The use of patient names often presents an ethical problem when confidentiality of information should be protected. When the patient's name is not available other unique identifiers should be selected. Medical insurance numbers and social security numbers, or even hospital record numbers are unique individual identifiers that are less ambiguous than the patient's name, are easily available, and could be used to link records on the same patient. A combination of medical insurance number, hospital record number, and date of admission was shown to be extremely accurate in linking patient records. Addition of the patient's name to these identifiers did not affect the accuracy of the linkage.[16]

In summary, record linkage should be based on a hierarchical procedure beginning with most unique identifiers and progressing to combinations of less unique identifiers. As an example the first run of record linkage may be based on medical insurance numbers. If all records are not matched a second run based on medical record number, gender and date of birth could be performed. The third run could involve the above variables with a part, e.g., first five letter of the patient's last name. This process could continue until a combination of all possible identifiers has been used. The risk of using a large number of identifiers is that records may not be matched if some of these identifiers are missing or misrecorded in one of the records.

When conducting record linkage three possible errors may occur. First, a record may not be matched with one from the other database(s). Second, a record may be matched with more than one record from the other database(s). Third, the record may be matched with a wrong record from the other database(s).

Detection of the two first errors may be accomplished with a simple manual review or using a computer algorithm. Detection of the third type of error, or record mismatch is not as straightforward. In order to detect these errors verification with at least one other identifier is essential. As an example for all records matched on medical insurance numbers, the dates of birth and gender could be compared. Mismatches or nonmatches should then be considered as potential errors and should be reviewed.

8. Statistical Uses of Registries

The data collected in a registry could be used to answer several questions ranging from simple description of changes in disease incidence or profile over time to more complex analyses evaluating the impact of treatment. The statistical questions that could be answered using the data from a registry are limited by the variable set, the resources available and the imagination of the investigators. In this respect the data from the registry could be used in the exact same way as data from any other observational study, although they may also subject to the same limitations of this study design.

One application of registry data that should be mentioned is the estimation of standardized rates of specific events. The patient population comprising large registries could be used as an external standard against which other smaller patient groups could be compared. This external standardization process produces estimates of expected rates of specific outcomes or events. These expected rates could be compared to the rates observed in the compared patient population or group. This comparison could then be used to address issues of differences in disease profile, evaluation of the quality of care or variation in treatment approaches.

To illustrate, the Major Trauma Outcome Study (MTOS) registry is one of the most widely used external standard population in the evaluation of quality of trauma care. The MTOS is a registry of major trauma patients treated at 100 level I trauma centers in the United States. Since these patients are treated at level I trauma centers it is assumed that they have received the best possible trauma care available in North America. The registry contains data on almost 100,000 patients collected over a period of 10 years. The data from the MTOS have been analyzed and logistic regression models (see Chapter 5) have been used to produce estimates of survival probabilities (P_S). These models are based on the patient's age, the Injury Severity Score, and the presence of penetrating trauma. The coefficients of the logistic regression model have been published and incorporated in the TRISS method for the evaluation of trauma care.[17]

Implementation of the TRISS method involves the following steps:

1. A cohort of patients with major trauma is assembled.
2. For each patient in the cohort P_S is calculated using the TRISS logistic regression.
3. The probability of death (P) is calculated as $P_D = 1 - P_S$ for each patient.
4. The sum of all P_S's in the cohort (%P_S) is equal to the number of expected deaths. (E)
5. The expected number of deaths is compared to that observed (0) by one of two methods:

   a) Flora's statistic:

   Flora's follows a normal distribution. Therefore if $Z \leq -1.96$ the observed mortality is significantly ($p < 0.05$) less than expected. If $Z \geq 1.96$ the observed morality is significantly ($p<0.05$) less than the expected.

   b) Standardized Mortality Ratio (SMR)

$$SMR = O/E$$

The SMR is an estimate of the Relative Risk for mortality associated with being in the compared cohort compared to the MTOS. Therefore if SMR > 1.0 the observed mortality is greater than the expected. Conversely if SMR < 1.0 the observed mortality is lower than expected. Confidence intervals around the point estimates of the SMR could be calculated as a measure of precision and statistical significance.[18]

Standardization based on registry data is not limited to mortality but with careful statistical considerations it could be expanded to other outcomes, or events, or use of specific interventions.

Conclusion

The use of registries in medical and surgical research is becoming increasingly popular and even more feasible in the recent years. The revolution of computer technology with increased capacity to manage large databases on personal computers and small local area networks have been pivotal in facilitating the development and use of these registries.

Implementation of a registry however requires careful planning that must take into consideration the availability of data, the requirements for stabling and maintaining the database and the ultimate use of the data. The process of data impute and analysis should be carefully mapped out and thoroughly tested. Finally, the issue of confidentiality should be given the utmost attention. Once these issues have been addressed disease specific registries could become a useful tool for research and quality assurance.

Selected Readings

1. Rutledge R. The goals, development, and use of trauma registries and trauma data sources in decision making in injury. Surgical Clinics of North America 1995; 75(2):305-326.
2. Ehlinger K, Gardner MJ, Nakayama DK. The trauma registry: An administrative and clinical tool. Topics in Health Record Management 1990; 11:43-48.
3. Mayer TA. Industrial models of continuous quality improvement: Implications for emergency medicine. Emergency Medicine Clinics of North America 1992; 10:523-547.
4. Lunge E. Implication for epidemiology of disease registers. Public Health Reviews 1993-1994; 21(3-4):263-270.
5. Klawansky S, Burdick E, Adams M et al. Use of the SEER Cancer Registry for technology assessment. International Journal of Technology Assessment in Health Care. 1991; 7(2):134-142.
6. Chancellor AM, Swingler RJ, Fraser H et al. Utility of Scottish morbidity and mortality data for epidemiology studies of motor neuron disease. Journal of Epidemiology and Community Health 1993; 47(2):116-120.
7. The Alpha$_1$-Antitrypsin Deficiency Registry Study Group. A registry of patients with severe deficiency of Alpha$_1$-Antitrypsin. Chest 1994; 106-1223-1232.
8. Borland International. Visual Dbase, 1995.
9. Taylor AJ. Objectvision 2.0. Bantam Books, Apr. 1992.
10. Noru_is MJ. SPSS. SPSS Publishing, 1999
11. Corel. Paradox 9, 1999.
12. Clarke DE. Development of a statewide trauma registry using multiple linked sources of data. Proceedings of the Annual Symposium on Computer Applications in Medical Care 1993:654-658.
13. Van Den Brandt PA, Schouten LJ, Goldbohm RA et al. Development of a record linkage protocol for use in the Dutch cancer registry for epidemiological research. International Journal of Epidemiology 1990; 19(3):553-558.
14. Keller JE, Howe HL, Noak JR. An algorithm for matching anonymous hospital discharge records used in occupational disease surveillance: Anonymous record matching algorithm. American Journal of Industrial Medicine 1991; 20:657-661.
15. Brenner H, Schmidtmann I, Stegmaier C. Effects of record linkage errors on registry-based follow-up studies. Statistics in Medicine 1997; 16:2633-2643.
16. Muse AG, Mikl J, Smith PE. Evaluating the quality of anonymous record linkage using deterministic procedures with the New York state aids registry and a hospital discharge file. Statistics in Medicine 1995; 14:499-509.
17. Champion HR, Copes WS, Sacco WJ et al. The Major Trauma Outcome Study: Establishing national norms for trauma care. Journal of Trauma 1990; 30(11):1356-65.
18. Sampalis JS, Lavoie A, Williams JI et al. Standardized mortality ratio analysis on a sample of severely injured patients from a large Canadian city without regionalised trauma care. Journal of Trauma 1992; 33(2):205-212.

CHAPTER 12

"Do-it-yourself Outcomes": Using Outcomes in Clinical Practice to Assess and Improve Quality of Care

Liane Feldman and Jeffrey Barkun

Evolution of the Outcomes Movement

A decade ago, Ellwood characterized the health care system as "an organism guided by misguided choices...the problem is our inability to measure and understand the effect of the choices of patients, payers and physicians on the patient's aspirations for a better quality of life".[1] The subsequent years have been marked by unprecedented growth in interest in measuring and understanding the outcomes of medical interventions. This phenomenon has been described as a "...revolution in medical care".[2] The ultimate goal of the outcomes movement is the improvement of quality of care through the assessment and analysis of the outcomes of health care.

Surgeons have long been interested in using outcomes to measure quality of care. At the turn of the century, Ernest Amory Codman, a prominent Boston surgeon and founder of the American College of Surgeons, promoted the idea that "the end result" of care, as assessed by long-term follow-up of patients, could be used to reflect the quality of care provided at different hospitals.[3,4] Eight decades later, the emphasis on outcomes and quality improvement has evolved from several converging streams.[5] First, the need for cost-containment stimulated efforts to identify ineffective interventions in order to eliminate unnecessary spending. As well, the cost-containment environment fostered the development of methods of measuring quality in order to ensure that it did not decline as costs were cut. Second, the documentation that significant regional variations existed in the use of medical procedures focused attention on the health outcomes associated with the different rates of utilization.[6] Third, the existence of large, computerized databases and of sophisticated data handling capacities enabled the analysis of aspects of care for literally hundreds of thousands of patients. Finally, the recognition that functionally-based outcomes and quality of life measures are important in judging the effectiveness of an intervention also contributed to the growth of the outcomes movement.

The growth of the modern outcomes movement has met with considerable concern from physicians. Many physicians are suspicious that the movement has more to do with cost-containment than with improving the health of their patients.[7] Physicians also fear the loss of control over the use of data generated from the care of their patients. Perhaps most concerning has been public reporting of outcomes data

Surgical Arithmetic: Epidemiological, Statistical and Outcome-Based Approach to Surgical Practice, by Lawrence Rosenberg, Lawrence Joseph, and Alan Barkun. ©2000 Landes Bioscience.

for individual surgeons, at times without reliable methods of accounting for the patients' comorbidities.[8-10] However, governments, managed care health plans and patient groups all support the implementation of methods to measure quality of care. In Canada, for example, the Health Minister has suggested grading the provincial health care systems, in a process styled after Maclean's magazine annual ranking of universities, using "information technology...to measure outcomes".[11] To ensure the appropriate use of such methods, physicians must be actively involved in the development and management of outcomes information systems.[12] Otherwise, physicians risk being judged by a technology with which they have little first-hand experience.[13]

Much of outcomes research has relied on large administrative databases, designed originally for billing purposes rather than patient care, to attempt to compare quality between hospitals. The principal outcome examined is often in-hospital mortality, which varies widely between institutions. Administrative databases are attractive because they contain large numbers of patients, are readily available, are computer readable and inexpensive to obtain.[14] However, insurance claims databases may lack relevant clinical information when compared with concurrently collected clinical data.[15] Claims databases often do not contain reliable information about patient factors that may confound outcome, such as comorbid disease and disease severity.[16] In order to assume that outcomes reflect quality across physicians or hospitals, patient factors that impact on quality must be accounted for in a process termed "risk-adjustment". Risk-adjustment is the critical task when comparing outcomes between individuals or institutions, and gaps in clinical information affects the validity of administrative data when used for quality of care comparisons.[14]

Rather than using administrative databases, several large studies have employed prospective clinical databases to calculate risk-adjusted outcomes. These large programs have then used these risk-adjusted outcomes to assess and attempt to improve quality of care in programs based on industrial models of "continuous quality improvement" (CQI).[17-19] While medicine is undoubtedly more complex than any industrial analogy can encompass, the concept of improvement through the collection and analysis of data is applicable. The literature now contains examples of very large programs of CQI, involving tens or even hundreds of thousands of patients, which will be discussed in further detail below. While most of us are not able to embark on such endeavors, the basic principle of using data to improve practice can be applied by individual surgeons or practice groups at the local level.[20,21] This Chapter will discuss how individuals can design a program in which outcomes data are used in everyday clinical practice to improve quality of care, an approach which can be termed "do-it-yourself outcomes". In essence, "do-it-yourself outcomes" involves the application of basic statistical principles to patient care with the goal of improving quality over time. However, it is not sufficient to merely record outcomes in clinical practice; using outcomes in a program of quality improvement involves the analysis of the processes of care to determine why the outcome, good or bad, occurred.

In order to demonstrate how to begin collecting and using outcomes in everyday surgical practice, we will first discuss quality improvement in general, define "outcomes", and examine how outcomes can be used to measure and improve quality of care. Examples from large published studies demonstrating the value of outcomes assessment for quality improvement will be discussed. Next, some of the pitfalls in the use of outcomes for quality improvement will be presented. We will then use data from a pilot project of outcomes collection to illustrate how this type of project can be implemented at the clinical level.[21]

Quality and Outcomes of Care: Definitions (Table 12.1)

Defining quality in medical care is an extremely complex task. Donabedian, the father of academic quality assessment, identified three aspects of quality: process, structure and outcome.[22] Until recently, the Joint Commission on Accreditation of Hospitals (JCAH) relied principally on structure and process criteria when assessing quality of care as part of its accreditation procedures for hospitals in the United States.[23] "Structure" includes the physical setting of the hospital or clinic, as well as the credentials of the health professionals. "Process" encompasses how services are provided to the patients moving through the system. In a simple example, a structural measurement would note whether cardiac monitors were present in the operating rooms, whereas a process measurement would note whether the cardiac monitors were actually used appropriately. Of course, neither structure nor process measure patient outcome directly. The underlying assumption of using process and structure criteria to judge the quality of care is that if these aspects are satisfactory, good outcomes will follow. In using process-based measures of quality for example, experts must first review the literature critically to create evidence-based practice guidelines, then judge whether patients received that standard of care. Unfortunately, some patients do not fit neatly into guidelines and other patients do not comply with the care provider's recommendations, which can cause measurement errors.[24] As well, it has been difficult to statistically correlate measures of process with measures of outcome. In one study where process and outcome were evaluated separately in a hypertension model and then compared, no significant association was detected between process and outcome.[25]

While structure and process criteria are often easier to define and measure, the pendulum has now swung to the use of outcome to assess quality in health care. Outcome refers to the patient's health status following a clinical encounter.[26] The concept of patient outcome is therefore as complex as the concept of "health". As Fraser has adroitly pointed out, the usual assessment of outcome in the postoperative clinic is the answer to the question "how are you doing" and the presence of a normally healing wound.[28]

Iezzoni has explained that using outcomes to answer questions about quality of care involves two steps.[27] The first step is to choose the outcomes of interest, while the second involves determining what caused the outcomes to occur. Traditionally, the outcomes of interest to surgeons have almost exclusively been mortality and morbidity. Clearly, the outcome of interest differs according to one's role in the health care system as patient, surgeon or payor. Thus, the outcomes of interest may be assessed from the patient's standpoint by measuring physical functional status, psychosocial functioning, role function (e.g., time off work), and quality of life. From the surgeon's standpoint, the focus might be on morbidity, mortality, physiologic stability, and the ease of using a new technique. Finally, from the payor's standpoint, one may examine the cost of care, readmission rate and length of hospital stay.[27] Patient-centered outcomes like functional status and quality of life may in

Table 12.1. Definition of terms

Quality of care: Degree to which health services for individuals and populations increase the likelihood of desired health outcomes and are consistent with current professional knowledge.[59]

Outcomes: Results that proceed from, or are at least associated with, the provision of health care services[46]

Continuous quality improvement: Industrial theory of quality improvement that is based on understanding and revising production processes based on information about the processes themselves, rather than focusing on individuals.[35]

Risk-adjustment: the process of accounting for relevant patient characteristics in order to draw inferences from outcomes about quality of care[27]

Health-related quality of life: The physical, psychological and social domains of health[51]

many cases provide more relevant measures of the intervention being studied than outcomes like death or acute complications.[28,29] For example, when studying inguinal hernia repairs, measuring the patient's convalescence time, amount of pain, and quality of life may give a fuller representation of a specific repair than only measuring the rate of wound infection or even recurrences.

There is a wide range of measurement tools available to assess concepts like health state and quality of life,[30] but few have been previously used to assess surgical patients. While it is important to use validated and reliable tools of measurement, instruments developed by epidemiologists and sociologists to assess quality of life may not have been designed to detect small, but clinically relevant, differences.[28,31] In addition to using a well-validated questionnaire to measure general health (e.g., SF-36),[32] an additional disease-specific questionnaire, either validated or ad hoc, should be used.[28,33] The disease-specific questionnaire should be adapted if possible from previous measures, rather than written from scratch, and should follow previously published guidelines.[31] Fraser has published a review of quality-of-life measurement in surgery.[28] Patient expectation and satisfaction are other important factors that one may wish to measure. Unfortunately, there is not much in the surgical literature addressing these concepts, and one may need to create one's own measures.

After defining the outcomes of interest, the second step involves determining what caused the outcomes to occur.[27] The determinants of patient outcome are multiple, complex and incompletely understood. The following relationship between patient factors, processes, structures and outcomes of care has been hypothesized:

processes+structures+patient factors+chance = outcomes.[12]

The patient's own attributes, including the severity of the illness being treated and the presence of other comorbid diseases, as well as the patient's motivation, education, and economic status, all may impact on the eventual outcome of treatment. It will seldom be possible to determine exactly how much of the outcome was due to the quality of care provided and how much was due to patient factors.[18] Outcomes of care are therefore not the final word on quality. Instead, outcomes are best used as Donabedian advised, "as cues that prompt and motivate the assessment of process and structure in a search for causes that can be remedied."[34]

Continuous Quality Improvement (CQI): Using Outcomes to Improve Quality

Traditional quality assessment in surgery is based on the identification and removal of "bad apples"–physicians whose performance fails to conform to implicit or explicit standards of care.[35] However, the natural response to being identified as the outlier on the low-end of the quality curve is to "become defensive, to distort data, and to shift responsibility elsewhere."[18] This approach clearly does not foster an environment conducive to true learning. In addition, searching for and removing the "bad doctors" underemphasizes the fact that physicians work in increasingly complex organizational systems, and errors can be made anywhere along in the process.[36] For example, a surgeon orders prophylactic antibiotics to prevent wound infection for a patient with appendicitis, thus following the guidelines for the medical care of this patient appropriately. However, in order for the patient to actually receive the medication, the prescription must be read and acted on by several different people (physicians, nurses, pharmacists, etc) in multiple departments. If the patient later develops a wound infection because the medication was not given properly, pointing only to the surgeon's responsibilities without addressing the other steps in the process will not improve the quality of the system.

In a complex system, therefore, the overall level of quality depends more on the system as a whole than on individual outliers.[37] As such, improvements in quality are more likely to result from small increments in the total system rather than from large changes by any individual within the system.[18,35] Berwick[35] and others[12,18,37] have suggested that this industrial model of "continuous quality improvement" is far better suited to the complex realities of modern health-care provision than the search for "bad apples". In the CQI approach, it is hypothesized that "real improvement in quality depends...on understanding and revising the production processes on the basis of data about the processes themselves."[35] The cornerstone of this model as applied to medicine is for physicians to have access to reliable information that
12 places the outcomes of their care in the context of their peers.[12] As Hammermeister notes, "monitoring outcomes of all health care episodes, as opposed to review of selected cases (e.g., adverse outcomes), has the advantage of greater statistical power, the opportunity to compare processes of care between good and bad outcomes, and the positive psychology of treating all providers equally."[18]

There are now several examples of how monitoring outcomes, coupled with the analysis of the processes and structures of care resulting in those outcomes, has been associated with improvements in quality of care. A classic example of how the compilation and ongoing feedback of outcomes of practice can be associated with improvements in quality comes from the study of wound infections. In 1965, Cruse, at the University of Calgary, began to prospectively record wound infections on the surgical services in a computer. Monthly and yearly, surgeons were provided with their rate of infection as well as the average rate in their division. Rigorous examination of the processes of surgical care associated with wound infection, including the timing of antibiotics, the role of shaving the wound, and multiple patient factors that impact on the likelihood of infection, were determined from Cruse's observational studies of tens of thousands of wounds. Beneficial effects were seen within six months of instituting the surveillance program. The approach provides an excellent example of the benefits of timely, nonadversarial outcomes collection and dissemination.[38]

Most evidence for the benefits of continuous quality improvement in surgery comes from cardiac surgery, where significant improvements in risk-adjusted mortality have been demonstrated in large outcomes-based quality improvement studies.[18,19,39] These studies have involved thousands of patients, scores of surgeons, and multiple institutions. In the Veterans Affairs (VA) Continuous Improvement in Cardiac Surgery Study (CICSS), nurse data coordinators prospectively abstracted 27 preoperative, 11 operative and 12 outcome variables from the patient chart onto paper forms or directly into a computer program at 44 medical centers (approximately 7000 cardiac surgery patients annually). Multivariate risk models were developed using univariate and logistic regression analyses of patients operated on in the first two years of the study. The ratio of observed (O)-to-expected (E) operative mortality (O/E ratio) was chosen as the primary indicator of quality for each center. The program directors of each center received data at six-month intervals, including a figure representing the O/E ratios for all centers with 90% confidence intervals. If risk-adjusted mortality exceeded expectations, audits of deaths were undertaken. During the 4.5 year period reported, the O/E mortality ratio for all centers decreased by 14%.[18]

The largest prospective outcome-based programs for quality improvement in noncardiac surgery was conducted by the National VA Surgical Quality Improvement Program (NS-QIP).[17] Between 1991 and 1997, data concerning 417 944 major surgical procedures were entered into the database. A multivariable risk-adjustment model was developed to identify high and low outlier centers by observed-to-expected outcome ratios. Feedback was provided annually to each facility enabling risk-adjusted comparison of outcomes with other hospitals. Low-outlier hospitals were invited to share the processes and structures of care they believed to have contributed to their results. High-outlier hospitals received suggestions about how to improve quality. Since 1994, the 30-day risk-adjusted morbidity has declined by 30%, and mortality has declined by 9%.[17] 12

Why these drops in risk-adjusted mortality and morbidity were seen in the VA projects is not known. The period over which the results are reported are several years long, which allows for multiple factors to be at play. The drop in mortality may have been from technologic advances in surgery unrelated to the quality improvement program. Because there were no control hospitals, in which data was collected but not relayed back to the surgical departments, one cannot assess whether improvements in quality occurred in the study period independent of the CQI program.

More convincing results for the benefits of CQI came from The Northern New England Cardiovascular Disease Study Group, where data were collected on 15 095 patients undergoing coronary artery bypass graft (CABG) surgery in five centers between 1987 and 1993. In 1990 and 1991, a three-component program aimed at reducing mortality was undertaken. The interventions included feedback of risk-adjusted outcomes data to clinicians three times each year, training in continuous quality improvement techniques, and round-robin site visits to observe the processes of CABG surgery at other centers.[39] Surgeons reported numerous changes in care that occurred as a result of the program.[40] During the postintervention period, there was a 24% reduction in the hospital mortality rate. In this study, the use of the preintervention period as the control group makes the conclusion that the CQI program was beneficial more convincing.

Although the above studies represent large-scale efforts, several smaller studies with more limited goals have also demonstrated the effectiveness of CQI programs. In one study where the goal was to reduce the rate of episiotomies, a CQI model focused on the processes of care by educating family practitioners about the indications for episiotomy. Following the CQI initiative, the episiotomy rate declined by 25% in almost 2500 women.[20] In a project described in detail below, we studied the rate of complications in general surgery patients before and immediately after the initiation of an outcomes-based program of quality improvement. We found that the overall rate of complications declined by almost 30% during the study period.[21] It has been noted by others that this first phase of improvement in clinical practice after the initiation of an outcomes-based program of CQI is usually easy, because obvious problems are identified and remedied.[41]

Limitations of Outcome Evaluation

Although the larger studies described above represent innovations in quality improvement, several limitations regarding outcome evaluation can be identified. These limitations stem in part from the large scale of the projects described above and the desire to compare outcomes between individuals or institutions. Some of these limitations can in fact be minimized by scaling down the project to the individual or group practice level while maintaining the same overall approach. Some of the pitfalls to watch for include:

1. What is the source of the data and who collects it? Data collection is often by data collectors reviewing charts.[12] Chart abstraction employs a secondary data source and is of unproven validity.[42,43] The medical record may not contain the desired information (particularly if functional measures are to be used), and is often incomplete or illegible. There are wide inter-observer variations in data like the physical exam. The data abstractor often does not know the patient or the medically relevant issues. The care providers themselves are dissociated from the data collection process, which is a cornerstone of the quality improvement program. This separation contributes to feelings of suspicion about the use of the data by regulatory bodies, rather than by clinicians. Because the surgeons do not control the data there is a time-lag before information can be used to aid clinical decision making. Additionally, data collection is expensive and redundant because it duplicates the medical record. The solution is the entry of patient data by the care-provider at the point of contact with the patient into a computerized record. This record is developed and managed by the surgeon and used both for patient care and outcome evaluation.[34]

2. Who measures the outcomes? When considering functional measures, the results can differ according to who does the measuring. For example, in our study comparing laparoscopic with open cholecystecotmy, the surgeons' estimates of duration of convalescence were 15% shorter than those of the study nurses'.[44] Physicians and patients also have divergent perceptions about the functional disabilities of patients.[45]

3. Whose patients are included? The involvement of a large number of surgeons complicates the ability to assess quality of care.[46] When many surgeons contribute data, and comparisons are to be made between them, the focus of the project's conclusions shifts from outcomes improvement to arguments over risk adjustment ("If my outcomes are worse it's because my patients are sicker- so you must not be adjusting for risk properly").[27] However, there are multiple patient-centered "risk" factors that may impact on outcome, including severity of illness, comorbidity, baseline functional status, expectations and preferences, and unidentified factors.[27] Therefore, a perfect system of risk-adjustment cannot exist. By scaling down the project and collecting data at the surgeon or group-practice level, where comparing outcomes between practitioners may be less important and the participants are involved in choosing the variables of interest, adjusting for case-mix can be simplified. For example, patients can be stratified by broad, easily defined and widely validated measures of preoperative risk (e.g., American Society of Anesthesiology (ASA) physical status category.[47] Expanding the project from a single surgeon to a practice group will increase the number of patients involved, allowing for more meaningful statistical analysis of the outcomes data.

4. What outcomes are measured? Morbidity and mortality are important endpoints, but functional outcomes are also important measures of outcome.[33,46,48] Both measures of general health, such as the SF-36[32,49] as well as disease-specific indices, can be included.[50,51] Obtaining information about the costs of the intervention under study may also be relevant. A good example of the use of multiple outcomes to assess a new procedure is in laparoscopic Nissen fundoplication. In addition to the usual morbidity and mortality data, investigators have compared physiologic scores (i.e., manometry and 24 hour pH) and symptom scores pre- and postoperatively to critically assess the operation.[52-54] Clearly, there is no single measure of outcome, and data from various sources (patient, record, physician) may need to be combined to judge outcome.

5. When are outcomes measured? The time needed to observe the outcome of an intervention is related to what outcome you wish to measure. While acute complications can be assessed during hospitalization, changes in quality of life must be measured over time (for example, at baseline, immediately postop, and at three months). Thus the ability of the principal surgeon to follow patients over the long-term after an intervention is extremely valuable.

6. How are the data organized? The focus of outcomes studies has been on procedures (e.g., coronary artery bypass grafting) rather than diseases or symptoms (e.g., coronary artery disease or angina).[12] This is probably because the type of procedure is readily available from administrative databases. As well, surgeons tend to organize their thinking about quality of care by procedure ("how many bile leaks have I had after laparoscopic cholecystectomy?"), while patients are more concerned with how their symptoms will be alleviated ("how will this operation relieve my biliary colic?"). Shifting the focus from a procedure to a symptom complex allows for the assessment of the appropriateness of various interventions for different patients and may be more relevant for the patient-physician interaction. In addition, entering patient data by symptom rather than procedure investigates the role of possible bias in procedure selection. Patients undergoing fundoplication procedures for GERD have

been selected by their surgeons. How do they differ from patients turned down for surgery or who decline surgery? In order to know if surgery is an effective treatment for GERD, one should not only compare patients pre-and postop, but compare them to patients receiving medical therapy.

In summary, with the wide variety of tools available for outcome assessment, it is best to remain pragmatic and measure variables that are relevant clinically. The science of using outcomes measurement to improve quality of care is nascent. Nonetheless, the key concept is to use outcomes "...as cues that prompt and motivate the assessment of process and structure in a search for causes that can be remedied".[55]

"Do It Yourself Outcomes" in Everyday Practice

There is no better way to obtain outcomes information than to collect data in the course of one's usual clinical practice. This allows the surgeon to be very close to the data. As discussed above, many of the limitations of outcomes studies are addressed when individuals begun to integrate tracking their own outcomes into their daily practice. In addition, this process eliminates the Orwellian feeling of having a regulatory body control patient data.

The first step in starting an outcomes program is to decide what preoperative, intraoperative and postoperative (outcomes) variables are to be collected. Consider a program where data about outcomes of cholecystectomy are of interest. Preoperatively, one might want to collect data about the patient's symptoms (biliary colic, dyspepsia, acute cholecystitis, pancreatitis, etc), the presence and severity of comorbid diseases (including the ASA class), laboratory and radiologic results, baseline quality of life, and patient expectations. Intraoperative variables could include the type of procedure performed (laparoscopic or open cholecystectomy), the operative findings, the duration of the procedure, and whether any complication occurred. Outcomes measures can be divided into three categories: those of interest to patients (e.g., satisfaction, quality of life, duration of convalescence), surgeons (e.g., bile duct injury, wound infection, retained stone), and payors (length of stay, readmission rates, costs).[56] The outcomes measures can thus include both those measured in the short-term (like the amount of postoperative pain or nausea) and longer-term (like the relief of symptoms or improved quality of life at three months postoperatively).

Next, the source of information is chosen. Ideally, the source is the patient at the time of each clinical encounter, and the data entered directly into a computer database in real time at the point of care. Questionnaires to assess quality of life should be administered at baseline and at regular intervals following the intervention of interest.

Information should be managed using a computer database (see Sampalis Chapter for a detailed discussion). There are several commercially available databases for outcomes assessment. However, they may not contain all the variables of interest. If the ready-made databases are not suitable, one can purchase a database program (e.g., Access or Filemaker Pro) and create one's own template, so that particular areas of interest can be evaluated as desired. One should also have software that can handle descriptive statistics and multivariable analysis.

Eventually, when data are collected on enough patients, multivariate analysis can be performed to tease out variables that may be predictive of the outcomes of interest, and to create a model to allow for risk-adjustment in order to determine when an unexpected outcomes has occurred (see Chapter 5). However, it is important to note that the process of collecting outcomes information can be useful well before predictive models can be developed. Any unusual outcome can be used to stimulate an examination of the process of care, which in itself may improve quality. One may begin to see patterns of outcomes that were not apparent prior to initiation of the project.

The American College of Surgeons Task Force on Outcomes recommends the following core-data set for surgery (Table 12.2). Disease-specific variables can be added as desired.[33]

Example: Pilot Project of Outcomes Assessment

In order to demonstrate the method and utility of outcomes collection at the individual practice level an example from a pilot project in a general surgery population, previously published,[21] will now be described. In this project, the goal was simply to establish and begin to use an outcomes database in a program of quality improvement. The principal outcomes followed were in-hospital morbidity and mortality. We compared the complications recorded in the database to those listed at morbidity and mortality rounds (M&Ms), a traditional cornerstone of surgical quality assurance programs. As well, we wanted to measure the impact, if any, of timely feedback of information about the rate of complications to the providers of care. This early database was not optimal in terms of the outcomes measured (provider biased, negative outcomes only), data abstraction (not by primary surgeon), and follow-up time (in-hospital only). Nonetheless, it does demonstrate the ease of beginning an outcomes program and the potential advantages inherent in such a project when it is designed, implemented and controlled by surgeons.

Methods

Database

Data were collected on consecutive patients operated on by one of three general surgery services at a McGill teaching hospital between October 1, 1995 and May 15, 1996. In addition to patients admitted from the emergency room, the clinical activities of the three staff surgeons on the service centered around laparoscopic, bariatric and pancreatic surgery, as well as renal transplantation. Data were collected prospectively by a surgical resident who was not directly involved in patient care. Information was obtained from the resident team caring for the patients, as well as from the chart.

Variable Selection

Preoperative, intraoperative and postoperative variables were chosen based on the National Veterans Administration Surgical Risk Study[57] and modified to reflect the patient population. The data were entered at the nursing stations onto a paper form then abstracted into a computer database created using FileMaker Pro 3.0©. Forty-eight preoperative (Table 12.3) and nine intraoperative (Table 12.4) variables were selected. Postoperatively, the database patients were followed throughout their hospital course and in-hospital adverse outcomes were recorded prospectively. Length of stay and readmission rates were also documented. Complications were graded

12

Table 12.2. Recommended data set for surgery (American College of Surgeons, with modification)[33]

Variables	Data Source
I. Pretreatment	
1. Clinical features	
Principal diagnosis	chart
Co-morbidities, age, sex	patient, chart
2. Nonclinical factors	
Health status	SF-36[32]
Patient preferences/expectations	instrument to be determined
II. Treatment	chart
III. Outcomes	
1. Clinical factors	
Survival	chart
Complications	chart
Disease-specific functional outcome measures	disease-specific questions as available
2. Nonclinical factors	
Costs	bill
Patient satisfaction	patient questionnaire
Health status	SF-36[32]

according to a modification of a classification scheme developed by Clavien et al.[58] In this classification, a complication is graded as I if it is minor and resolves spontaneously if left untreated, II if potentially life-threatening and requires interventions that have inherent risk (Iia is managed with medications, while Iib requires further invasive therapy), III if the complication is associated with residual disability or organ loss, and IV if the complication results in mortality.

12

Feedback of Outcomes

The outcomes recorded in the database were reported back to the three staff surgeons and the chief residents of the service at three month intervals. As well, the results were presented twice to the entire department over the length of the project.

Results

Database

311 patients underwent 325 operations in the study period. Their average age was 50.9 ± 17.1 years (range 16 to 92); 44.4% were female. The ten most frequent preoperative risk factors (Table 12.3) were smoking (25.6%), hypertension (18.3%), loss of greater than 10% of body weight (6.4%), shortness of breath on exertion (6.1%), psychiatric disorder requiring medication (4.5%), obesity (4.2%), dialysis (4.2%), alcohol consumption of greater than two drinks per day (3.5%), and chronic oral corticosteroid use (3.2%). No preoperative risk factor was identified in 33.5% of the patients, whereas 7.4% had four or more.

The five most common operative procedures were cholecystectomy (63), inguinal hernia repair (54), gastric bypass (36), partial colectomy (29), and appendectomy

Table 12.3. Frequency of preoperative variables (n =311 patients). Modified from Feldman L, Barkun J, Barkun A, et al. Surgery 1997; 122:711-20

Risk Factor	Frequency	Percent
Smoking within 2 weeks of index procedure	80	25.6
Hypertension requiring medication	57	18.3
History of asthma	31	10
More than 10% body weight loss in previous 6 months	20	6.4
Dyspnea on exertion	19	6.1
Psychiatric history on medication	14	4.5
Obesity (body mass index >35)	13	4.2
Currently on renal dialysis	13*	4.2*
Diabetes mellitus (insulin dependent)	13	4.2
Alcohol consumption more than 2 drinks per day	11	3.5
Currently taking oral corticosteroids	10	3.2
Previous operation with cardiopulmonary bypass	9	2.9
Diabetes mellitus (noninsulin dependent)	9	2.9
Previous deep venous thrombosis	8	2.6
Disseminated carcinoma	8	2.6
Sleep apnea on continuous positive airway pressure	8	2.6
Status postoperative revascularization/amputation	8	2.6
Bleeding disorder (including coumadin use)	7	2.3
History of previous cerebrovascular accident	6	1.9
Angina within previous 1 month	6	1.9
Angina more than 1 month prior to surgery	4	1.3
History of prostatism	4	1.3
Currently in atrial fibrillation	3	1
Patient is "do not resuscitate" status	3	1
Transfusion greater than 4 units of blood this admission	3	1
Open wound or wound infection	2	0.6
History of severe chronic obstructive pulmonary disease	2	0.6
History of percutaneous coronary angioplasty	2	0.6
Cirrhosis	2	0.6
Functional health status: fully dependent	2	0.6
Dyspnea at rest	1	0.3
Dyspnea on minimal exertion	1	0.3
Ventilator dependent	1	0.3
Congestive heart failure more than 1 month prior to surgery	1	0.3
Functional health status: partially dependent	1	0.3
Paraplegia	1	0.3
History of previous transient ischemic attack	1	0.3
Drug addiction	0	0
Acute renal failure	0	0
Ascites	0	0
Esophageal varices	0	0
Active hepatitis	0	0
Myocardial infarction within previous 12 months	0	0
Myocardial infarction within previous 3 months	0	0
Congestive heart failure within previous 1 month	0	0
Quadriplegia	0	0
Hemiplegia	0	0
Central nervous system tumor	0	0

*12/13 patients were for renal transplantation

12

Table 12.4. Frequency of intraoperative variables (n = 325 index procedures). Modified from Feldman L, Barkun J, Barkun A, et al. Surgery 1997; 122:711-20.

Variable	Frequency	Percent
Emergency room admission	78	24
Preoperative hospital stay		
0 days	122	37.5
1 day	141	43.4
2-5 days	39	12
>5 days	23	7.1
Principal anesthesia technique		
General	310	95.4
Epidural or spinal	10	3.1
Local	5	1.5
ASA class		
Not available	18	5.5
1	102	31.4
2	130	40.0
3	70	21.5
4	5	1.5
5	0	0
Primary surgeon		
Staff	20	6.2
PGY1	24	7.4
PGY2	30	9.2
PGY3	39	12
PGY4	31	9.5
PGY5	181	55.7
Operating time, hours		
< 1	124	38.2
1 to < 2	107	32.9
2 to < 3	65	20.0
3 to < 4	17	5.2
4 to < 5	6	1.9
> 5	4	1.2
Not available	2	0.6
Intraoperative blood loss		
Not available	127	39.1
0 – 20 ml	44	13.5
21 to 100 ml	55	16.9
101 to 500 ml	80	24.6
501 to 1000 ml	15	4.6
>1000 ml	4	1.2
Units of blood transfused intraoperatively		
0	315	96.9
1	4	1.2
2 or more	6	1.9

(17). 68.6% of the procedures followed elective admissions. General anesthesia was used in 95.4% of the cases. 23% of the patients were designated as ASA class 3 or 4.

The mean length of stay was 9.3 ± 13 days (median 6 days; range 0 to 152 days). Patients who experienced an adverse outcome had a mean length of stay of 17.2 ± 19.4 days, whereas those without complication had a mean length of stay of 5.2 ±5.5 days (95% CI for difference in mean length of stay 8.2 to 15.8 days). Readmission for complications within 30 days of the index procedure occurred in 27 patients (8.9%). The most common reason for readmission was a wound infection (8 patients, 30%).

There were 179 adverse postoperative outcomes in 106 patients (34.1%). The in-hospital mortality rate was 2.6%. Grading of the complications revealed that 46.4% were grade I, 23.5% grade IIa, 21.8% grade IIb, 3.9% grade III, and 4.5% grade IV. On univariate analysis, the occurrence of a postoperative complication was associated with an ASA grade of three or four, the presence of any preoperative risk factor, and an operation which lasted more than two hours.

Presentation of Complications at M&M Conference

Of the 106 patients who experienced postoperative complications in the database, 29 were presented at M&Ms (27.4%; 95% CI 19% to 36%). The percentage of database patients with complications presented at M&Ms increased with the grade of their most severe complication. Whereas 15.4% (95% CI 6% to 25%) of database patients with grade I complications were presented, this proportion increased to 30% for grades II and III combined (95% CI 17% to 44%), and 87.5% for grade IV(95% CI 65% to 100%).

Effect of Feedback on Complications

The rate of complications declined from the first half of the project to the second half. Whereas 58 of 142 admissions in the first part of the study led to complications (41%; 95% CI 33% to 49%), 53 of 183 admissions (29%; 95% CI 22% to 36%) in the second part led to adverse outcomes (95% CI for difference in complications from part 1 to part 2: 1.4% to 22.2%). Grading of the complications revealed similar proportions of minor and severe complications in the first and second halves. While the overall rate of complications declined in the second part of the study, the patient population remained stable with regard to age, ASA, number of preoperative comorbid factors, and length of stay.

In one instance, interim analysis of the database revealed an unexpectedly high rate of wound complications in patients undergoing bariatric surgery (six instances in 15 cases, 40%; 95% CI 16-68%). This was relayed to the attending surgeon, and changes in the process of care, including a change in antibiotic type and method of drain placement, were made. In the 21 subsequent cases, only two wound complications were noted (9.5%; 95% CI 1-27%; 95% CI for decrease in wound complications: 3-58%).

Conclusions

The pilot project described above demonstrates how a homegrown outcomes project can be designed and implemented with relative ease. Data collection can become a seamless part of the daily routine of caring for patients. The prospective recording and analysis of preoperative, intraoperative and postoperative outcomes may improve one's results if unexpected outcomes are used as flags to examine the process of care.[12]

The nature of the database allowed for comparison of changes in practice over time in a manner not usually possible with traditional, peer-review methods of quality assessment like the M&M conference. As demonstrated in the pilot project, the M&M conference concentrated on the most serious complications but only picked up 15.4% of the "minor" complications. While not life-threatening, these minor complications nonetheless are associated with significantly increased lengths of stay, and presumably have an impact on patient well-being and satisfaction (although this was not addressed in the pilot study).

Reduction of unwanted variance in the system of care provision can be more effective in improving overall quality than an emphasis on punishing individuals ("bad apples").[37] In fact, we recorded a significant decline in complications in the second part of the project, largely attributable to improvements in care made in one group of patients. The recording of data allowed for a pattern of adverse outcomes, previously unrecognized using peer-review methods of quality assessment, to emerge and be addressed. The fact that data continued to be collected after changes were implemented allowed for the impact of the changes to be recorded.

Methodological issues that plague the large outcomes projects are more easily addressed when the project is at the individual surgeon or group-practice level. When the goal of the project is to use the data to aid in everyday decision making and to improve personal quality, the issue of risk-adjustment, which is so critical when comparisons are made between individuals, becomes far less important. When a sufficient number of patients are entered into the database, multivariable analysis can then be used to create an appropriate risk-adjustment model. Data collection can be done by the care-provider at the point of contact with the patient, directly into the office or clinic desktop computer. Questionnaires assessing general and disease-specific health can be given to the patient while he/she is waiting to be seen. The follow-up time in the database continues for as long as the practitioner follows the patient. In essence, the database is the medical record in statistically analyzable form.

Surgeons have always used data to judge the risk of a procedure on an individual and this fundamental role is enhanced by the analysis of outcomes. By proactively implementing outcomes data collection and analysis, the feeling that a regulatory body controls patient decision making is largely alleviated. As Ebert has noted, "fortunately or unfortunately, the day when one may answer a clinical question from a patient by saying 'in my experience' or 'in my opinion this is the best method of treatment' has probably been relegated to history."[41] Using outcomes information to help answer patients' questions about effective treatments is the way of the future. As surgeons, we therefore should continue to be involved in the design and implementation of outcomes projects in our everyday practice.

Selected Readings

1. Ellwood PM. Shattuck lecture-Outcomes management. A Technology of patient experience. NEJM 1988; 318(23):1549-1556.
2. Relman AS. Assessment and accountability. The third revolution in health care. NEJM 1988; 319(18):1220-1222.

3. Neuhauser D. Ernest Amory Codman, M.D., and end results of medical care. Int J Tech Assess in Health Care 1990; 6:307-325.
4. Campion FX, Rosenblatt MS. Quality assurance and medical outcomes in the era of cost containment. Surg Clin N Amer 1996; 76(1):139-159.
5. Epstein AM. The outcomes movement- Will it get us where we want to go? NEJM 1990; 323(4):266-270.
6. Wennberg JE, Mulley AG, Hanley D et al. An assessment of prostatectomy for benign urinary tract obstruction. Geographic variations and the evaluation of medical care outcomes. JAMA 1988; 259(20):3027-3030.
7. McAleese P, Odling-Smee W. The effect of complications on length of stay. Ann Surg 1994; 220:740-4.
8. Choudhry NK, Wright JG, Singer PA. Outcome rates for individual surgeons: Concerns about accuracy, completeness, and consequences of disclosure. Surgery 1994; 115(3):406-408.
9. Chassin MR, Hannan EL, Debuono BA. Benefits and hazards of reporting medical outcomes publicly. NEJM 1996; 334(6):393-398.
10. Griffith BP, Hattler BG, Hardesty RL et al. The need for accurate risk-adjusted measures of outcome in surgery. Lessons learned through coronary artery bypass. Ann Surg 1995; 222(4):593-599.
11. Rock backs away from health care report card. News Tickers. www.canoe.ca, 1999.
12. Hammermeister KE. Participatory continuous improvement. Ann Thor Surg 1994; 58(1815-21).
13. Blumenthal D, Epstein AM. Quality of health care. Part 6: The role of physicians in the future of quality management. NEJM 1996; 335(17):1328-1331.
14. Iezzoni LI. Assessing quality using administrative data. Ann Intern Med 1997; 127:666-674.
15. Jollis JG, Ancukiewicz M, DeLong ER et al. Discordance of databases designed for claims payment versus clinical information systems. Implications for outcomes research. Ann Intern Med 1993; 119:844-50.
16. Wen SW, Hernandez R, Naylor D. Pitfalls in nonrandomized outcomes studies. The case of incidental appendectomy with open cholecystectomy. JAMA 1995; 274:1687-1691.
17. Khuri SF, Daley J, Henderson W et al. The Department of Veterans Affairs' NSQIP. The first national, validated, outcome-based, risk-adjusted, and peer-controlled program for the measurement and enhancement of the quality of surgical care. Ann Surg 1998; 228(4):491-507.
18. Hammermeister KE, Johnson R, Marshall G, Grover FL. Continuous assessment and improvement in quality of care. A model from the Department of Veterans Affairs Cardiac Surgery. Ann Surg 1994; 219(3):281-90.
19. Hannan EL, Kilburn H, Racz M et al. Improving the outcomes of coronary artery bypass surgery in New York state. JAMA 1994; 271(10):761-766.
20. Reynolds JL. Reducing the frequency of episiotomies through a continuous quality improvement program. Can Med Assoc J 1995; 153(3):275-282.
21. Feldman L, Barkun J, Barkun A et al. Measuring postoperative complications in general surgery patients using an outcomes-based strategy: comparison with complications presented at morbidity and mortality rounds. Surgery 1997; 122:711-20
22. Donabedian A. The quality of care. How can it be assessed? JAMA 1988; 260(12):1743-1748.
23. Schroeder SA. Outcome assessment 70 years later: are we ready. NEJM 1987; 316(3):160-162.

12

24. Palmer Rh. Process-based measures of quality: The need for detailed clinical data in large health care databases. Ann Intern Med 1997; 127:733-738.
25. Nobrega FT, Morrow GW, Smoldt RK, Offord KP. Quality assessment in hypertension: analysis of process and outcome methods. NEJM 1977; 296:145-8.
26. Brook RH, McGlynn EA, Cleary PD. Quality of health care. Part 2: Measuring quality of care. NEJM 1996; 335(13):966-970.
27. Iezzoni LI. Using risk-adjusted outcomes to assess clinical practice: an overview of issues pertaining to risk adjustment. Ann Thor Surg 1994; 58(1822-6).
28. Fraser SCA. Quality-of-life measurement in surgical practice. Br J Surg 1993; 80(2):163-169.
29. Kreder HJ, Wright JG, McLeod R. Outcomes studies in clinical research. Surgery 1997; 121(2):223-225.
30. Patrick DL, Deyo RA. Generic and disease-specific measures in assessing health status and quality of life. Med Care 1989; 27:S217-32.
31. Guyatt GH, Bombardier C, Tugwell PX. Measuring disease-specific quality of life in clinical trials. Can Med Assoc J 1986; 134:889-95.
32. Ware JEJ, Sherbourne CD. The MOS 36-item short-form health survey. I. Conceptual framework and item selection. Med Care 1992; 30:473-83.
33. Reemtsma K, Morgan M. Outcomes assessment: a primer. Bull Amer Coll Surg 1997:34-9.
34. Hammermeister KE, Daley J, Grover F. Using outcomes data to improve clinical practice: what we have learned. Ann Thor Surg 1994; 58:1809-11.
35. Berwick DM. Continuous improvement as an ideal in health care. N Engl J Med 1989; 320(1):53-56.
36. Lafell G, Blumenthal D. The case for using industrial quality management science in health care organizations. JAMA 1989; 262(20):2869-2873.
37. Kritchevsky SB, Simmons BP. Continuous quality improvement. Concepts and applications for physician care. JAMA 1991; 266(13):1817-1823.
38. Cruse PJE, Foord R. The epidemiology of wound infection. A 10 year prospective study of 62, 939 wounds. Surg Clin North Am 1980; 60:27-40.
39. O'Connor GT, Plume SK, Olmstead EM et al. A regional intervention to improve hospital mortality associated with coronary artery bypass graft surgery. JAMA 1996; 275:841-846.
40. Nugent WC, Schults WC. Playing by the numbers: how collecting out outcomes data changed my life. Ann Thor Surg 1994; 58:1866-70.
41. Ebert PA. The importance of data in improving practice: effective clinical use of outcomes data. Ann Thor Surg 1994; 58:1812-4.
42. Brennan TA, Localio AR, Leape LL et al. Identification of adverse events occurring during hospitalization: a cross-sectional study of litigation, quality assurance, and medical records at two teaching hospitals. Ann Int Med 1990; 112:221-6.
43. O'Neil AC, Petersen LA, Cook EF et al. Physician reporting compared with medical-record review to identify adverse medical events. Ann Int Med 1993; 119:370-6.
44. Barkun JS, Barkun AN, Sampalis JS et al. Randomized controlled trial of laparoscopic versus mini cholecystectomy. Lancet 1992; 340:1116-19.
45. Calkins D, Rubenstein LV, Cleary PD et al. Failure of physicians to recognize functional disability in ambulatory patients. Ann Int Med 1991; 114:451-454.
46. Harrigan M. Quality of care: issues and challenges in the 90's. A literature review. Stittsville, Ontario: Canadian Medical Association, 1992.
47. Scher KS, Scott-Conner CEH. making peer review statistically accountable. Am J Surg 1996; 171:441-444.
48. Ikegami N. Functional assessment and its place in health care. NEJM 1995; 332(9):598-99.

49. McHorney CA, Ware JEJ, Rachel JF, Sherbourne CD. The MOS 36-item short-form health survey (SF-36):III. Tests of data quality, scaling assumptions, and reliability across diverse patient groups. Med Care 1994; 32:40-66.
50. Wright JG, McLeod RS, Lossing A et al. Measurement in surgical clinical research. Surgery 1996; 191(3):241-244.
51. Testa MA, Simonson DC. Assessment of quality-of-life outcomes. NEJM 1996; 334(13):835-840.
52. Watson DI, Jamieson GC, Pike GK et al. Prospective randomized double-blind trial between laparoscopic Nissen Fundoplication and anterior parital fundoplication. Br J Surg 1999; 86:123-130.
53. Trus TL, Laycock WS, Saring JP et al. Improvement in quality of life measures after laparoscopic antireflux surgery. Ann Surg 1999; 229:331-336.
54. Feldman LS, Mayround S, Antiniuk M, Fried GM. Patient-centered outcomes of laparoscopic Nissen Fundoplication for gastroesophageal reflux disease. CJS 1999; 42:A133.
55. Donabedian A. Quality assessment and assurance: unity of purpose, diversity of means. Inquiry 1988; 25:173-92.
56. Birkmeyer JD. Outcomes research and surgeons. Surgery 1998; 124(3):477-83.
57. Khuri SF, Daley J, Henderson W et al. The National Veterans Administration Surgical Risk Study: Risk adjustment for the comparative assessment of the quality of surgical care. J Am Coll Surg 1995; 180:519-528.
58. Clavien PA, Sanabria JR, Strasberg SM. Proposed classification of complications of surgery with examples of utility in cholecystectomy. Surgery 1992; 111:518-26.
59. Lohr KN, Donaldson MS, Harris-Wehling J. Medicare: a strategy for quality assurance. V. Quality of care in a changing health care environment. Qual Rev Bull 1992; 18:120-6.

12

CHAPTER 13

Technology Assessment

Jeffrey Barkun and Alan Barkun

In the preceding Chapters of this book, tools have been put forth that are essential to the clinician scientist to carry out proper clinical evaluations and make clinical decisions. These have included reviews of basic methodological concepts in epidemiology and biostatistics, as well as step-by-step discussions regarding different methodologies and issues, including diagnostic tests, clinical trials, costing, and pharmaco-economic analysis and decision trees. One ultimate goal is to assess any existing or emerging health care technology. Used in its broadest sense, this activity ranges from the evaluation of how to carry out a history and physical examination to using an innovative imaging method, or delivering state of the art therapy with new medications or innovative surgical equipment. We will now try and put the process of technology assessment into context while applying some of the concepts acquired in the book to date.

Introduction

Technology assessment is a paradigm whereby new technology needs to be thoroughly assessed from the time of its first development in vitro or in an animal model until the time of its widespread acceptance in the routine medical care of patients. Clinicians may be most familiar with parts of this process in the context of the introduction of a new drug and the phases I to IV of drug trial development.[1] One great similarity lies in the vulnerability that both eye-catching technology and promising new drugs both may be prematurely popularized amidst overwhelming patient and media pressure. The introduction of a new technology, however, involves additional and unique evaluative aspects related to such specific issues as operator dependency, the existence of a "learning curve" phenomenon, the amortized cost of equipment and its maintenance, and the need for operator accreditation, to name but a few. Pertinent examples are the new endoscopic and laparoscopic techniques that have emerged and revolutionized Gastroenterology and Surgery. These have been introduced in spite of ever decreasing health care budgets, rendering their proper assessment essential.[2,3] It is therefore regrettable that provincial and national research budgets have only sporadically recognized the usefulness of such evaluative research.

This Chapter will attempt to highlight central issues in technology assessment while using examples related to the assessment of novel laparoscopic and endoscopic procedures.

Surgical Arithmetic: Epidemiological, Statistical and Outcome-Based Approach to Surgical Practice, by Lawrence Rosenberg, Lawrence Joseph, and Alan Barkun. ©2000 Landes Bioscience.

The Need for Technology Assessment

It is tempting to attribute the triumph of medical practice over a given disease to a specific intervention or technique but more often than not, simultaneous advances in overall patient care rather than the introduction of a specific new technology may be responsible in large part for this improvement. This is one reason why assessment of a new technology is necessary, preferably in the form of a comparative trial where the control group is concurrent rather than historical, thus appropriately reflecting the "current standard of care". On the other hand, not every new technology can or should undergo a formal comparative assessment. For example, it would be unethical to carry out a randomized controlled trial comparing colonoscopic polypectomy to its surgical counterpart.[4] Such potentially large differences in patient outcomes are, however, infrequent.

It has become increasingly clear over the past decades that the quality of health care delivery is not a simple function of the level of technological sophistication applied, and that novel technology is nearly always associated with greater direct costs. The onus is therefore on the clinician scientist to demonstrate the added value for clinical practice provided by the adoption of a new technology (or its lack). As discussed in the Chapter on decision analysis, the priority of clinician scientists should remain that of effectiveness (i.e., quality of care) over costs. Nonetheless, the generation of costing and effectiveness data can be viewed as the balancing act of Technology assessment. Ultimately, both health technology assessment as a whole and health care economics in particular attempt to provide information so that three key factors can be balanced: access to health care (equity), quality of health care (effectiveness), and cost or cost-efficiency of health care provision.[5] This exercise leads to the concept of performing technological trade-offs, which can be addressed from numerous viewpoints: political, administrative, and ethical.[6] These trade-offs are often socially and politically sensitive, and include equity among age groups, social classes, concept of need, and legitimacy of therapeutic goal. Some of these will be further developed in the following sections.

Models of Technology Assessment

Different models of technological change have been proposed. Although an exhaustive discussion about their characteristics is beyond the scope of this Chapter, two concepts characterize the recent trends in technology assessment.

First, is the broadening of the classical "reductionistic biomechanical model" of endpoint measurement.[7] Indeed, there are more than just abnormalities in biochemistry, physiology, and gene structure or regulation that need to be assessed when evaluating a new technology. The additional measures relate to the fields of clinical epidemiology and social psychology and include less traditional but validated and objective measures such as quality of life and patient satisfaction.[7] This latter so-called "hermeneutic model" is based on a critical paradigm applied implicitly in every day delivery of patient care, and strongly influences the design of trials evaluating novel technologies. The second important concept is the explicit recognition that technology assessment is no longer a simple linear process linking basic research to clinical testing with its subsequent diffusion into everyday practice. Models initially developed amidst this simplified paradigm viewed innovation as driven by scientific progress ("science push"), or induced by market demand ("demand pull").[8] Yet it is now recognized that the process of technology assessment has become infinitely more sophisticated and includes a multitude of political and social influences that may explain some of its inherent shortcomings.[8] Examples include the symbiotic nature of interests of medical specialties and Industry, the growing importance of patient perspectives (through advocacy groups, for example), and heterogeneity in socio-cultural perspectives across different countries.

Regardless of the conceptual model adopted, the ultimate success of a technology assessment exercise will lie in the successful diffusion of worthwhile technology, and the effective condemnation of just another "gimmick".[9]

The Types of Technologies to Be Assessed

The technologies to be assessed may be diagnostic, such as the role of Magnetic Resonance Cholangiography (MRCP) in evaluating a jaundiced patient[10] or endoscopic ultrasound in detecting and staging gastrointestinal malignancies,[11,12] and may also apply to methods of screening for the prevention of disease such as colonoscopy in the screening for colonic polyps.[13] More commonly, however, the technology to be assessed is therapeutic, such as the performance of laparoscopic inguinal hernia repair.[14] In its broad sense, the evaluation may also apply to already commonly practiced procedures such as determining the predictors of a "helpful" gastroscopy (in the sense of influencing management) rather than just a "positive" gastroscopy (in the sense of abnormal endoscopic findings),[15] or the use of a standard catheter versus a sphincterotome in achieving common bile duct cannulation at endoscopic retrograde cholangio-pancreatography.[16]

As a rule, the assessment of new devices, for example those used in novel laparo-endoscopic techniques, has lagged behind the performance of pharmacological trials. One reason for this discrepancy relates to differing regulatory requirements.[17,18] Additional factors might include a differing institutional structure within which development decision making takes place. Indeed, in the case of surgical procedures, such as in the initial development and popularization of laparoscopy, the process is often carried out by physicians in clinical practice as opposed to corporate, academic, and government clinical research settings which is usually the case for the development of new drugs. The development of new operations or devices must also consider the element of manual skill, and an historical inability to measure components of surgical techniques as precisely as a chemical composition, or a physical structure. As a probable reflection of these differences, comparative evaluations in the surgical literature have been, in general, less frequent than in medical reporting,[20,21] and their number did not increase significantly from 1980 to 1990.[22] Although this trend is now clearly reversing, the quality of reported comparative trials has recently been questioned.[23] In particular, the quality of economic evaluations has not always been constant. For example, in a review of economic evaluations on knee arthroplasty available from 1966-1996, none of 40 studies met established criteria to form a comprehensive economic evaluation.[24] In order to reverse this trend, there have been pleas for more extensive economic input at the early stages of trial design.[25]

The Development of a Technology–The Life Cycle of Technology Assessment

Technology assessment can be viewed as an integrative exercise aimed at establishing state of the art knowledge of a specific technology with a view to shaping policy recommendations concerning its adoption and utilization in practice both in the short- and long-terms.[26] It is hence the necessary link between health science and health policy.[27] This is why technology assessment not only focuses on clinical outcomes but also on the evaluation of the benefits and the costs (clinical, societal, and system-wide) of transferring the technology to clinical practice.[28] The development of a technology from its inception to its adoption in clinical practice is comprised of a series of evolutionary stages, which both alter the technology itself and reshape the health care system it is designed to improve. This dynamic process adopts different perspectives, as the technique becomes more widely used. It should be seen as a continuous loop, which helps to refine the technology and fine tune its application. Many different steps are involved in the "life cycle" of a technology as it is developed and assessed. Investigators have reported several levels of evaluation common to the development of diagnostic or therapeutic technologies as disparate as medical imaging, surgical practice, drugs, and picture archiving and communication systems.[29,30] In general terms, there are primary technology assessments through which new data are generated including feasibility studies, epidemiological observational studies as well as trials yielding data on efficacy and effectiveness including randomized controlled trials and medical effectiveness studies.[31] Secondary health technology assessments make use of existing data and include such methods as cost-effectiveness and cost-benefit analyses, computer modeling, systematic literature synthesis and meta-analysis, as well as ethical, legal, and social assessments.[31] Examples of these different assessment steps are listed below. The tools for the evaluation of these steps have been described, for the most part, in the preceding Chapters and will not be discussed in great detail here.

Feasibility

Early on, laboratory and, usually, animal studies will establish technical feasibility and preliminary data on the safety of a technique. It is then applied to small or medium-sized series of selected patients thus yielding preliminary results based on observational studies. Unfortunately, new techniques are often unjustifiably adopted at this stage without further evaluation, even though efficacy and effectiveness have yet to be assessed. What's more, premature adoption of the technology at this point often represents a missed opportunity for truly objective evaluation. Laparoscopic cholecystectomy thus was already heralded as the gold standard in treating cholelithiasis prior to formal large scale comparisons with "mini" cholecystectomy even though certain patient outcomes were later found to be similar.[32]

Efficacy and Effectiveness

These two properties refer to the measured benefits of a treatment when compared with pre-existing standards of care (which may include no specific treatment). Efficacy refers to the performance of a technology when measured under ideal circumstances: ideal patients, expert technicians, perfectly clear and appropriate indications, etc...-thus limiting the external validity or generalizability of the trial. Effectiveness relates to the application of the novel technology when the assessment is performed under real life conditions that would usually include a more heterogeneous group of patients and varied operator expertise.[33] This discrepancy arises because an "efficacy trial" attempts to control most extraneous variables that may influence final outcome by restricting them in order to assess as "purely" as possible the impact of the technology studied. In contrast, an effectiveness trial will above all attempt to recreate real life with all its imperfections.

Cost Effectiveness Analyses

Once effectiveness has been established, cost-benefit and effectiveness analyses determine the efficiency of a technique by linking benefits to cost. Such analyses may be included as part of controlled trials, but were part of randomized clinical trials in only 0.2% of over 50,000 such studies published between 1966 and 1988,[34] although the proportion is now increasing. Most often, however, they are carried out as a form of secondary technology assessment where the effectiveness component is derived from existing data in the literature (such as in decision modeling). It is often at this stage of the evaluative process that the assessment moves from an individual to a societal perspective (cost-effectiveness or cost-benefit analyses can also view aggregate costs from other perspectives, including the provider's, or the insurer's). The unit of interest is at this stage no longer an individual patient, but rather a group of patients undergoing the technology. As advocates of individual patients, clinicians have questioned the relevance of this frame of reference to clinical practice.[35] In fact, cost-effectiveness analyses have had relatively low impact on practice for several reasons: there often is an imbalance between attention to cost and attention to outcome (e.g., quality of care). Also, most health care providers have a short-term parochial financial perspective whereas cost-effectiveness analyses often take a long-term view that captures all costs, benefits and hazards.[36] There also has been poor standardization of methodology, and unrealistic expectations that fundamental ethical and political questions could be answered. Mostly though, there has been a failure of Society to accept the need for allocating scarce resources more judiciously.[37] There has been a lack of decision-making on the part of payers and government because of the difficulty in agreeing on the numerous trade-offs to be considered.[38]

Outcomes Research

At the later stages of technological development, methods of assessment usually employ large data bases,[39] practice variation data, and decision modeling techniques—all with the not always implicit goal of developing practice guidelines or clinical management tools. Monitoring systems such as quality assurance and maintenance programs[40] provide clinicians with feedback and education and aim to modify behavior, and improve guidelines. A typical example is the assessment of regional practice variations with respect to a procedure: for e.g., cholecystectomy or hysterectomy. Outcomes research can therefore be seen to complete the cycle of technology assessment[41] by providing feedback once it has been widely accepted and used. Please refer to the Chapter by Feldman and Barkun in this text for a more in-depth discussion of outcomes research.

Nonclinical Assessments

The need for additional assessments is increasingly recognized and includes ethical, legal, and social evaluations.[42] This poses significant methodological problems, and requires a range of methods, both quantitative and qualitative, as well as a different set of skills from those required for clinical evaluation. It has been suggested that much can be learned from interviewing relevant people including those receiving treatment, their relatives, friends, and those providing care.[31] As an example of this type of analysis, there has been a recent plea to carry out more structured "error analyses", despite their possible legal consequences, in the hope of further improving surgical care provided with new technologies.[43] In the mind of the authors of this Chapter, and others,[31,43] such evaluations should also fall into the general realm of health technology assessment in its broad sense.

We will now consider issues that are more specifically related to the assessment of emerging technologies:

The Timing of Technology Assessment

There exists no widely accepted formula to determine the optimal timing of a technology assessment. However, most authors agree that such evaluations should be carried out at an early stage of their development.[31] Some have even suggested that early identification of emerging technologies be submitted to a type of "early warning system" that would rely on the review of early scientific and clinical literature, and on the opinions of well-informed senior professionals, before more formal classical technology assessment results are obtained. Such a system may provide policy makers with much needed lead time to be able to develop policies before their implementation becomes too difficult or impossible.[8,31] However, such a system may also marginalize different views on the desirability of a given technology. Examples of this type of "early warning" process include some of the activities of the Canadian Coordinating Office for Health Technology assessment[44] and more specifically for novel surgical procedures, The Senate of the Royal Surgical Colleges of Great Britain and Ireland.[45] The latter group has also proposed to oversee the development and defining of training and credentialling requirements for these new technologies.

Some authors, in contrast, have suggested that the evaluation should only take place after the technology has been allowed to diffuse into routine practice, i.e., when true effectiveness is known.[46] The rationale for this dissenting view is that the technology must already be adequately developed to reflect a close to optimal level of effectiveness. Moreover, it needs to generate sufficient interest to justify time and expense while maintaining patient and investigator motivation.

Not uncommonly, the rapid development of new technologies may antiquate the results of well-designed and performed clinical trials, even before completion or widespread diffusion of their results. This has been the case for medical treatments of cholelithiasis, such as extracorporeal shock wave lithotripsy, with the advent of laparoscopic cholecystectomy,[47,48] and the assessment of gastroesophageal reflux disease drug therapy prior to the omeprazole era.[49] Some groups have thus suggested a critical point at which the assessment of an emerging technology should be carried out[50] with the ideal window alternately being determined as akin to the beginning of a feasibility study[51] or the time at which opinion leaders of the innovating group are still hesitant to adopt the new technology.[52] Regardless of its actual effectiveness, once a technology is perceived as useful, particularly with the aide of high profile media support, it may become very difficult to encourage appropriate evaluation due to pressure from physicians and patients.

Learning Curve

A critically limiting issue for the timing of the evaluation of a new device which is especially true for physician operators, such as surgeons and endoscopists, is the "learning curve" phenomenon. It requires that participating operators have acquired enough experience to master the technique effectively prior to beginning a study.[53,54] Any premature assessment could underestimate the existence of a "learning curve" and bias results by underestimating the true performance of a new technique. However, the timing of the evaluated technology cannot be such that it has already replaced the previous gold standard.

Operator Variability

As was the case for generalizability in speaking of patient selection, the expertise sought amongst participating surgeons or endoscopists in formal trials may limit the generalizability of results to centers with differing expertise owing to operator selection. For example, the ERCP cannulation success rate of 80-95% described in studies treating patients with early acute pancreatitis may be greater than what is the widely available endoscopic skill in this context.[55-58] Varying operator expertise may become a problem in the performance of multi-center trials although it can be argued that this heterogeneity in performance is a reflection of the "real life" situation. In order to obviate this problem and optimize the use of existing expertise, Van der Linden has suggested randomizing patients to a given surgeon rather than to a treatment group.[59]

Iterative Technology Assessment

As mentioned above, initial adoption of a technology may be only the beginning of an often prolonged iterative process in which important redesigning takes place based on user feedback. This process can then be viewed as a review of policy against a backdrop of health technology assessment rather than a one-time study.[50,60] The multiple assessments would then occur at different times in the life of that technology. Any significant change in circumstance that might affect the given technology might warrant a review of its assessment; however the basis for the identification of such times is undefined, and the potential for bias in choosing such periods remains a possibility.

Diffusion of Study Results

"Technology is only as effective as the clinicians who employ it and the system in which they function".[61] This is perhaps why experience has shown that there is not a direct relationship between the quality of a given technology evaluation and the subsequent popularity of its results in health care delivery ("diffusion"). This is not surprising when one considers that there is often no formal way in which technology evaluation results are reported, applied, or taught. In fact, many factors other than the conclusions of a technology evaluation will primarily influence the impact of study conclusions on clinical practice. These include societal pressures, remuneration considerations, actual comprehension of study conclusions, and parochialism. What's more, different factors may be more important in different geographical areas or physician groups. Occasionally, the poor diffusion of a novel technology may be traced back to an assessment where the "significant" conclusions were based on sole statistical considerations rather than clinical relevance (see Chapter 2 for a discussion of the important distinction between clinical and statistical significance) . Such seems to have been the case for laparoscopic appendectomy where postoperative wound infection, according to a recent meta-analysis, seems to have been the only meaningful clinical benefit over open appendectomy.[62] It is increasingly clear that physician education alone is probably not sufficient to bring about changes in clinical practice be it from a costing or quality of care point of view.[63,64] In fact, probably as important as the quality of the trials generated in the phases of technology assessment are the ability to change clinical practice, and the ability to establish a lasting change in clinical practice. In effect, there is a significant difference between providing information and imparting knowledge.[65]

Several studies have looked at the effects of simply providing information to clinicians in order to affect their behavior: for example, the use of price stickers on anesthetic supplies has not been shown to change clinical behavior with respect to cost-effectiveness.[66] What's more, if a change can be demonstrated, its short duration brings into question whether it is only an observational effect related to the study itself, similar to a "Hawthorne effect".[67] It has become increasingly clear that an intervention is needed to supplement physician information in order to bring about a lasting change in behavior. Interventions have thus used techniques such as "performance feedback" to help decrease length of hospital stay by providing rates of appropriate hospital stay for preceding time periods.[68] When one is targeting housestaff, behavior modification seems to be particularly sensitive to the presence of an interested physician leader.[69] Another strategy has been the use of physician reminders[70] or even to embed information into order sheets. For example, laboratory and radiological test utilization was found to decrease when physicians had to check off relevant clinical indications on a test order sheet.[71] However, when a booklet providing the information alone was handed out, omitting the order form, no change occurred. Perhaps, the most important factor to trigger lasting change is the influence of leadership, and the perception that change is part of a group process, although the targeted group needs to exhibit fairly close interaction.[72]

It is hoped that learning theory and management theory may eventually help to bring about effective strategies to alter physician behavior, particularly with respect to much maligned costing information.

Special Issues in Health Technology Assessment

Technology Assessment and Clinical Practice Guidelines

As part of the Evidence Based Medicine movement, the 1990s produced outcomes research and practice guidelines, many of which have stemmed from or impacted on selected aspects of technology assessment. The Agency for Health Care Policy and Research (AHCPR) in the USA has been one of the leading organizations in the production of clinical practice guidelines. Mandated by the American

Congress to address issues of clinical variability, the AHCPR has generated clinical practice guidelines using multidisciplinary panels that performed comprehensive, systematic reviews of the data, meta analyses (if advisable and appropriate), and clinical consensus when little data were available to support previous guideline statements.[73] These have led to successful implementations of selected guidelines, especially when local adaptation of these guidelines has involved patients, providers, payers and purchasers. AHCPR has been involved in the National Guideline Clearinghouse project in collaboration with the American Association of Health Plans (an umbrella organization for a number of managed care organizations), and the American Medical Association which are all cosponsoring a Web-based repository for clinical practice guidelines and related materials. The Web site for this initiative is <www.guideline.gov>. The AHCPR and the National Guidelines Clearinghouse are discussed in this Chapter as an example of possible new approaches to technology assessment diffusion and evaluation. They are also included as a source of reference for interested readers. This type of initiative will likely increase in popularity with time.

The Status of Health Technology Assessment Worldwide

The results of an international survey completed in 1995 and subsequently published in 1997 included the participation of 103 organizations from 24 countries.[74] This survey, in addition to listing established organizations active in health technology assessment surveyed the types of organizations, their research priorities and agenda, the types of technologies assessed, the stages in the life cycle of technologies assessed, the attributes of a technology that are evaluated, the assessment methods used, the method of dissemination employed, and identified the primary users of these data. In addition to providing further substantive information to the reader, this compendium may be a useful reference tool.

Ethical Issues in Technology Assessment

Since the 1970s, two approaches have dominated the analysis of ethical issues in health care.[75] The "deontological" framework is based on concepts of duty and required actions. In this concept, persons are respected as ends in themselves, and are never treated as means to an end. Its categorical imperative is that moral judgment be rational, free of self-interest and universalizable. The "teleological" framework is based on the outcome of actions and the value of particular endpoints. Teleological analysis is typically phrased in terms of the value of the consequences of specific choices and deeds, and it is often exemplified by the utilitarian maxim that moral actions are those that result in the greatest benefit for the greatest number, defined subjectively by those who benefit. The applicability of such concepts to the science of technology assessment is self-evident. Irrespective of the framework for analysis, ethical questions in technology assessment can be grouped into five broad, interrelated categories. They include issues related to essential concepts and definitions, to diagnosis, to prevention and therapy, to research, and to allocation. Some of these are alluded to in a number of paragraphs above, and the reader is referred to a recent authoritative review for more details.[75] It is likely that, as time progresses, such considerations will, rightfully, increasingly influence developments in the science of technology assessment.

Conclusion

The science of technology assessment faces new challenges because of the current multitudes of technologies to be evaluated, and because of the advent of entirely new types of technologies such as simulation studies for broad usage of telecooperation techniques,[76] the assessment of health care information systems,[77-79] and the creation of new partnerships for e.g., with health care insurers.[80] Some authors have even suggested that professional societies at this time should focus more on developing guidelines for technology assessment itself rather than on guidelines for technology utilization.[81] Clinical trials and technology assessment are increasingly difficult to design and carry out, but their performance is an integral part of our duties as clinicians in order to identify and answer meaningful clinical questions and subsequently deliver optimal care. Amidst an era of ever increasing budgetary constraints, payers seek objective and valid scientific data to make difficult, but hopefully enlightened decisions about rationalization of resources. Clinicians must thus possess a sound understanding of study methodologies in order to maintain major roles in the interpretation and performance of technology assessment.

Selected Readings

1. Jadad AR. Types of randomized controlled trials. In: Randomized Controlled Trials–A User's Guide. London: BMJ Publishing Group. 1998: 14-16.
2. Cotton PB. Interventional gastroenterology (endoscopy) at the crossroads: A plea for restructuring in digestive diseases. Gastroenterology 1994; 107:294-299.
3. American College of Physicians. Access to health care. Ann Intern Med 1990; 112:641-661.
4. Cotton PB. Therapeutic gastrointestinal endoscopy: Problems in proving efficacy. N Engl J Med 1992; 326:1626-1628.
5. Horl WH, de Alvaro F, Williams PF. Healthcare systems and end-stage renal disease (ESRD) therapies—an international review: Access to ESRD treatments. [Review]. Nephrology, Dialysis, Transplantation. 1999; 14 Suppl 6:10-5.
6. Giacomini MK. The which-hunt: assembling health technologies for assessment and rationing [see comments]. [Review]. J Health Politics, Policy & Law. 1999 Aug; 24(4):715-58.
7. Lorenz W, Troidl H, Solomkin JS, Nies C et al. Second Step: Testing – Outcome Measurements. World J Surg 1999; 23:768-80.
8. Blume S. Early warning in the light of theories of technological change. Internat J Technol Assess Healthcare. 1998; 14(4):613-23.
9. Traverso LW. Technology and surgery: Dilemma of the gimmick, true advances, and cost effectiveness. Surg Clin N Amer Feb 1996; 76(1):129-138.
10. Barkun JS, Barkun AN. Jaundice. In: Scientific American-Surgery. 1997, CDROM.
11. Palazzo L. Imaging and staging of bilio-pancreatic tumours: role of endoscopic and intraductal ultrasonography and guided cytology. Ann Oncol 1999; 10 Suppl 4:25-7
12. Woolfolk GM, Wiersema MJ. Endosonography in the assessment of esophageal stenosis. Gastroint Endos Clin N Amer 1998 Apr; 8(2):315-28.
13. Bond J. Guidelines for the screening and follow-up of patients with colonic polyps. Am J Gastroenterol. (in press):2000.
14. Barkun JS. Length of stay and cost and effectiveness of laparoscopic herniorraphy. J. Am Coll Surg. 2000(editorial)(in press).
15. Naji SA, Brunt PW, Hagen S et al. Improving the selection of patients for upper gastrointestinal endoscopy. Gut 1993; 34:187-91.

16. Cortas GA, Mehta SN, Abraham NS et al. Selective cannulation of the common bile duct: a prospective randomized trial comparing standard catheters to sphincterotomes. Gastrointest Endosc. 1999; 50:775-9.
17. American College of Physicians. Access to health care. Ann Intern Med 1990; 112:641-661.
18. FDA: Technology assessment committee. Device evaluation and the Food and Drug Administration Process. Am Soc Gastrointest Endosc, July 1994.
19. Torres WE, Adwers J, Baumgartner BR et al. Extracorporeal shock-wave biliary lithotripsy. N Engl J Med 1991; 324:1813-4.
20. Colditz GA, Miller NJ, Mosteller F. How study design affects outcomes in comparisons of therapy. I and II: Medical. Statistics In Medicine 1989; 8:441-454:455-466.
21. Gray DT, Hewitt P, Chalmers TC. The evaluation of surgical therapy. In: Rutkow IM, ed. The Socioeconomics of Surgical Health Care Delivery. St. Louis, Missouri: CV Mosby Co 1989:228-56.
22. Solomon MJ, McLeod RS. Clinical studies in surgical journals- Have we improved? Dis Colon Rectum 1993; 36:43-48.
23. Bell PR. Surgical research and randomized trials. Br J Surg. 1997; 84:737-8.
24, Saleh KJ, Fagni A, Macaulay WB et al. Understanding economic evaluations: A review of the knee arthroplasty literature. American Journal of Knee Surgery. 1999 Summer; 12(3):155-60.
25. Sculpher M, Drummond M, Buxton M. Economic evaluation in health care research and development: Undertake it early and often. Uxbridge: Health Economics Research Group, Brunel University, 1995.
26. Luce BR, Brown RE. The use of technology assessment by hospitals, health maintenance organizations, and third-party-payers in the United States. Int J Technolog Assess Health Care. 1995; 11:79-92.
27. Barkun JS, Barkun AN, Mulder DS et al. Technology Assessment. In: Principles and practice of research: Strategies for surgical investigators. 2nd ed. Springer-Verlag, 1991: 313-321.
28. Russell I. Can it work? Does it work? Research design for health technology assessment. York: University of York, 1996.
29. Banta HD, Luce BR. Health care technology and its assessment: An international perspective. Oxford: Oxford University Press, 1993.
30. Fryback DG, Thornbury JR. The efficacy of diagnostic imaging. Med Decis Making. 1991; 11:88-94.
31. Mowatt G, Bower DJ, Brebner JA et al. Int J Tech Assess Health Care. 1998; 14(2):372-386.
32. Lorenz W, Troidl H, Solomkin JS et al. Second Step: Testing – Outcome Measurements. World J Surg 1999; 23:768-80.
33. Bond JH. Outcomes and effectiveness of endoscopic procedures. Gastrointest Endosc 1992; 38:725-727.
34. Adams ME, McCall NT, Gray DT et al. Economic analysis in randomized controlled trials. Medical Care 1992; 30:231-43.
35. Asch DA, Hershey JC. Why some health policies don't make sense at the bedside. Ann Int Med 1995; 122:846-50.
36. Berger ML. The once and future application of cost-effectiveness analysis. Joint Commission Journal on Quality Improvement. 1999 Sep; 25(9):455-61.
37. Berger ML. The once and future application of cost-effectiveness analysis. Journal on Quality Improvement. 1999 Sep; 25(9):455-461.
38. Giacomini MK. The which-hunt: assembling health technologies for assessment and rationing [see comments]. [Review]. Journal of Health Politics, Policy & Law 1999 Aug; 24(4):715-58.

39. Steiner CA, Bass EB, Talamini MA et al. Surgical rates and operative mortality for open and laparoscopic cholecystectomy in Maryland. N Engl J Med 1994; 330:403-8.
40. Sapienza PE, Levine GM, Pomerantz S et al. Impact of a quality assurance program on gastrointestinal endoscopy. Gastroenterology 1992; 102:387-393.
41. Bouchard S, Barkun AN, Barkun JS et al. Technology assessment in laparoscopic general surgery and gastrointestinal endoscopy—science or convenience? Gastroenterology 1996; 110:915-925.
42. Dolan A, Zingg W. Health care technology: How can we tell if we can afford it? A Canadian viewpoint. J Long term Effects of Med Implants 1993; 3:277-82.
43. Troidl H. Disasters of endoscopic surgery and how to avoid them: Error analysis. World J Sur 1999; 23:846-55.
44. Battista RN, Feeny DH, Hodge MJ. Evaluation of the Canadian Coordinating Office for Health Technology Assessment. Int J Technolog Assess Health Care. 1995; 11:102-16.
45. Border P. Minimal access ("keyhole") surgery and its implications. London: Parliamentary Office of Science and Technology, 1995.
46. McGregor M. Can our health services be saved by technology evaluation? The Quebec experience. Clin Invest Med 1994; 17:334-42.
47. Barkun AN, Barkun JS, Sampalis JS et al and the McGill Gallstone Treatment Group. A randomized clinical trial comparing gallbladder stone shock wave lithotripsy to laparoscopic cholecystectomy. Gastroenterology 1993; 104(4):A2.
48. Nicholl JP, Brazier JE, Milner PC et al. Randomized controlled trial of cost-effectiveness of lithotripsy and open cholecystectomy as treatments for gallbladder stones. Lancet 1992; 340:801-7.
49. Spechler SJ, and the Dept. of Veterans Affairs Gastroesophageal Reflux Disease Study Group. Comparison of medical and surgical therapy for complicated gastroesophageal reflux disease in veterans. N Engl J Med 1992; 326:786-92.
50. Banta HD, Andreasen PB. The political dimension in health care technology assessment programs. Int J Technology Assess Health Care. 1990; 6:115-23.
51. Bunker JP, Hinkley D, McDermott WV. Surgical innovation and its evaluation. Science 1978; 200:937-41.
52. Stocking B. Factors influencing the effectiveness of mechanisms to control medical technology. In: B. Stocking ed. Expensive Health Technologies. Oxford: Oxford University Press, 1988.
53. Sigman HH, Fried GM, Hinchey J et al. Role of the teaching hospital in the development of a laparoscopic cholecystectomy program. Can J Surg 1992; 35:49-54.
54. Cass OW, Freeman ML, Peine CJ et al. Objective evaluation of endoscopy skills during training. Ann Intern Med 1993; 118:40-44.
55. Fan ST, Lai ECS, Mok FPT et al. Early treatment of acute biliary pancreatitis by endoscopic papillotomy. N Engl J Med 1993; 328:228-32.
56. Williamson RCN. Endoscopic sphincterotomy in the early treatment of acute pancreatitis. N Engl J Med 1993; 328:279-280.
57. Neoptolemos JP, London NJ, James D et al. Controlled trial of urgent endoscopic retrograde cholangiopancreatography and endoscopic sphincterotomy versus conservative treatment for acute pancreatitis due to gallstones. Lancet 1988, October 29:979-983.
58. Neoptolemos JP, Carr-Locke DL, London N et al. ERCP findings and the role of endoscopic sphincterotomy in acute gallstone pancreatitis. Br J Surg 1988; 75:954-60.

13

59. Willem van der Linden. Pitfalls in randomized surgical trials. Surgery 1980; 87:258-262.
60. Banta HD, Thacker SB. The case for reassessment of health care technology: Once is not enough. JAMA 1990; 264:235-40.
61. Ahrens T. Impact of technology on costs and patient outcome. Crit Care Nurs Clin N Am. 1998; 10:117-25.
62. Fingerhut A, Millat B, Borrie F. Laparoscopic versus open appendectomy: time to decide. World J Surg 1999 Aug 23; (8):835-45.
63. Eisenberg JM. Doctors' decisions and the cost of medical care: The reasons for doctors' practice patterns and ways to change them. Ann Arbor, MI: Health Administration Press, 1986.
64. Greco PJ, Eisenberg JM. Changing physicians' practices. N Eng J Med 1993; 329:1271-74.
65. Lowe PF, Eisenberg JM. Can information on cost improve clinicians' behavior? In J Tech Assess Health Care. 1997; 13(4):553-61.
66. Horrow JC, Rosenberg H. Price stickers do not alter drug usage. Can J Anaest 1994; 41:1047-52.
67. Tierney WM, Miller ME, McDonald CJ. The effect on test ordering of informing physicians of the charges for outpatient diagnostic tests. N Eng J Med 1990; 322:1499-504.
68. Studnicki J, Stevens CE, Knisely L. Impact of a cybernetic system of feedback to physicians on inappropriate hospital use. J Med Edu 1985; 60:454-60.
69. Billi JE, Duran-Arenas L, Wise CG et al. The effects of a low-cost intervention program on hospital costs. J Gen Inter Med 1992; 7:411-17.
70. Frazier LM, Brown JT, Divine GW et al. Can physician education lower the cost of prescription drugs? Ann Int Med 1991; 115:116-21.
71. Durand-Zaleski I, Rymer JC, Roudot-Thoravai F et al. Reducing unnecessary laboratory use with new test request form: Example of tumor markers. Lancet. 1993; 342:150-3.
72. Schectman J, Kanwal N, Schroth W et al. The effect of an education and feedback intervention on group-model and network-model health maintenance organization physician prescribing behavior. Medical Care 1995; 33:134-44.
73. Pearson KC. Role of evidence-based medicine and clinical practice guidelines in treatment decisions. Clin Ther 1998; 20(suppl.):C80-85..
74. Perry S, Gardner E, Thamer M. The status of health technology assessment worldwide. Results of an international survey. Int J Technolog Assess Health Care 1997; 13:81-9)
75. Heitman E. Ethical issues in technology assessment. Conceptual categories and procedural considerations. Int J Technlog Assess Health Care 1998; 14:544-66.
76. Kumbruck C, Schneider MJ. Simulation studies: a new method of prospective technology assessment and design. Quality of Life Research 1999; 8(1-2):161-70.
77. Cushman R. Serious technology assessment for health care information technology. J Amer Med Infor Assoc 1997 Jul-Aug; 4(4):259-65.
78. Anderson JG, Aydin CE. Evaluating the impact of health care information systems. Int J Tech Assess Health Care. 1997 Spring; 13(2)380-93.
79. Langlotz CP, Seshadri S. Technology assessment methods for radiology systems. Radiol Clin N Amer 1996 May; 34(3):667-79.
80. Ballard GT. Medical and dental technology assessment. Technology and Health Care. 1996 Sep; 4(3):291-303.
81. Lagasse RS. Monitoring and analysis of outcome studies. Internat Anesthesiol Clin 1996 Summer; 34(3):263-77.

Index

D

E

F

G

H

I

K

L

M

N

O

P

Q

R

S

T

U

V

W

For Product Safety Concerns and Information please contact our EU representative GPSR@taylorandfrancis.com Taylor & Francis Verlag GmbH, Kaufingerstraße 24, 80331 München, Germany

Printed and bound by CPI Group (UK) Ltd, Croydon, CR0 4YY
22/09/2026
02221138-0001